GW01454286

To Rosie,

Always wanted
to see you. Thanks for
your enthusiasm for my

books!

All the very best,

Brad Barkley

Brad Borkan and David Hirzel

it takes two

or three

The Superpower of Small Teams

From Hollywood to the Moon & Everything In Between

ISBN 978-1-945312-21-2

Cover design by Anne Sharples

TERRA NOVA PRESS

Terra Nova Press
P. O. Box 1808
Pacifica CA 94044

Also by Brad Borkan and David Hirzel

When Your Life Depends on It:
Extreme Decision Making Lessons from the Antarctic

Audacious Goals, Remarkable Results:
*How an Explorer, an Engineer and a Statesman
Shaped our Modern World*

Dedication

Brad:
To my brothers, Gary and Ron.
It has been the privilege of a lifetime to have
been part of our very small team. Our parents,
Harold & Jean, would be pleased.

David:
To the teams that made the team that made me:
My grandparents, Edna & Elezear A. Hirzel,
and Eva & Irwin M. Bender.

★

Contents

"A Gilbert is of no use without a Sullivan"

—W.S. Gilbert

Prologue

We Are A Team Now

We three wouldn't be here if we hadn't steered our lives this way, right? Fighter pilot, test pilot, astronaut – risky jobs, those. We're the individuals whose experience in probing the unknown will show the way for everyone who's going to come after. Now where do we go next?

As high as we can – in our case, already halfway to the moon.

A risky plan at best, but everything that could possibly be imagined had been considered to ensure our safety and that of the mission. And to reassure our families.

Then, Bang!

What if something never imagined were in fact possible, so possible that from the moment our spacecraft lurched – right at that moment, the moment of the bang – we knew something bad had happened. The gauges on the sensors to our absolute most precious oxygen and hydrogen tanks began to plummet – followed by those showing the strength of the command module's batteries, and for all we could tell at that moment, the length our lives.

Little by little we get the real picture. Mission Control confirms that our long-awaited moon landing will never be. And now we need to tweak our course just a little to keep us in the return-to-Earth orbit.

But there's a catch.

We may or may not have enough fuel left to us to burn for that course adjustment. Suppose the burn fails in any of the possible ways it can – because ignition didn't happen exactly right, or because the captain left his thumb on the throttle too long, or not long enough. Or something else we haven't even considered.

And if that happens, we will slip out of the return-to-Earth orbit and into one that goes around the sun, and we three in our space capsule will become yet another asteroid, cast into a never-ending orbit.

We wait for the signal from Mission Control. We three are each specialists in our fields – the commander, the navigator, and the lunar orbiter. In the original plan, two of us would have descended to walk on the moon's dusty surface while the third waited alone in the space capsule – momentarily becoming the loneliest, most remote man in the universe, until the lunar lander rises up to rejoin the orbiter and our team is reunited for the trip home to Earth.

In an unexpected turn of events, we three were pulled together four days before launch from the best of the specialists in each of those three fields. We had been mere acquaintances before that, high-achievers in each of our chosen fields, but in our training members of other – different – three-man teams.

Not anymore.

Now we are in this hopelessly damaged, slowly dying spaceship together, doing our best to finish the mission we have long trained for. If any of us feels fear over the prospect of what will happen to us if the burn fails, we will not reveal it. We're all professionals here. Together we wait for that first critical burn, the one that will shape our course and determine our fates, if not forever then for a few more hours.

Until the next critical burn will have to happen, at its specific moment, for its specific time, under the thumb of our commander.

In this we are together.

We are now a team. A very small team.

★

This was a real event faced by three men. What happened next is described later in this book.

Introduction

The Need To Collaborate

Throughout time, very small teams have worked together to advance society. They have shaped our world and built our future.

Teams of two or three people have come together in endeavors ranging from exploration and discovery to science and invention, and even in such disparate fields as art and design, music, pop culture and comedy. Real, and even fictional, very small teams have been glorified in Hollywood films, in television shows and children's cartoons.

In almost every undertaking – ranging from Susan B. Anthony and Elizabeth Cady Stanton's tireless work in the 1800s to establish women's rights to Katalin Kariko and Drew Weissman creating a vaccine for Covid for which they won a Nobel Prize – very small teams have played an enormous role.

One reason for this is that while the human mind works exceedingly well at an individual level, in some cases and for some individuals, it works even better when joining forces with one or two other like-minded people. That's when the real power kicks in.

Their singular strengths multiply and amplify when working in the capacity of a very small team. The team members can share their abilities, ideas and dreams, willing to dare mighty things while overcoming obstacles, and possibly changing the world in small and big ways as they progress along their journey.

As our societies become more complex, we grow ever more

dependent on finding creative solutions to deal with tough problems. In today's world we are faced with challenges that could not have been imagined by previous generations. The solutions, if they exist at all, will be derived from past experiences and raw knowledge. Seldom does one person have enough of either to fully solve a complex problem on their own. A pair or trio of people working together can comprehend and achieve more than the same two or three people each working alone.

It's easy to call that 'teamwork,' but it is much more than that. It is camaraderie and collaboration toward a common goal.

The Amazing Power of the Very Small Team

Famous very small teams are all around us. As you will soon see, there are exciting stories to be told about decision-making within the limits of teams of two and three people.

How did they come together? How did they share the creative principal, and why did they choose one direction over another? What makes a team successful, and why do some successful teams fall out at the height of their fame?

These are questions we have set out to answer.

Consider Wilbur and Orville Wright, two brothers whose shared intellect and ambition helped them to solve the twin mysteries of powered, heavier-than-air flight and how to control a flying machine. Each brought their own insights and inspirations to their work. Together they pursued and refined that work, building the foundations of air travel as we know and use it today.

But have you heard of Charles Taylor, the man who created the lightweight, water-cooled gasoline engine that turned the Wright brothers' glider into the first powered, propeller-driven airplane? How did Taylor get involved with the brothers, and why did this trio succeed when large corporations and governments could not?

Wherever we looked we found high-profile pairs and trios who shaped our modern world, and whose stories are filled with little-known facts that make each story even more fun to share.

There are plenty of famous pairs of people like Ben Cohen and

Jerry Greenfield the creators of Ben and Jerry's ice cream. What would life be like without ice creams named Cherry Garcia and Chunky Monkey? And the famous two Steves – Steve Jobs and Steve Wozniak – who, together, created the first Apple computer, which laid the groundwork for today's ubiquitous iPhone. What about the candle maker William Procter and soap maker James Gamble, who, together, created the corporate giant of Procter and Gamble? They were brothers-in-law whose father-in-law suggested the two men partner together, simply because candles and soap used the same raw materials.

The entertainment industry is brimming with much-loved pairings like Fred Astaire and Ginger Rogers. Where would musical theatre be today without Rodgers and Hammerstein? For that matter, where would modern culture be without the satire and wit of Gilbert and Sullivan's comic operas? And there are many more contemporary examples, like the magicians Penn and Teller, and the Hollywood fictional duo Thelma and Louise.

Writers of these stories

The more we, as a pair of authors, thought about our own teamwork – working in two different countries separated by eight time zones, knowing that together we had already co-created two books, both better than either of us could have ever crafted on our own – the more we realized our story is not unique. In fact, that gave us an edge in thinking about very small teams, because we are one.

We all have a need to collaborate at various points in our lives and small teams are how we do it. Think about your own lives. Each of you, our readers, is already a member of a small team. You were born into a family; perhaps you've started one of your own. You may have a cluster of close friends; you may have joined a book club or a recreational sports team. Perhaps you enjoy the company of certain work colleagues, or hang out with a group of like-minded people who enjoy the same pursuits as you.

What about love? We all are born with a primal urge to find a mate and forge a life together – the fundamental, intimate team of two navigating through the complexities of modern life. And why do we have best friends – that one person who knows us inside out?

Very small teams can bring great satisfaction and yield remarkable results. They can also harbor tremendous stresses. The angst suffered by some of the teams in this book was very, very real.

What, exactly, is a team?

When we first settled on very small teams as a general idea, huge realms of exciting investigation began to unfold before us. We narrowed our focus to teams of two, at most three, whose energies and intellects combined to find a solution to a problem, who followed a path toward a result that others could only dream about, who created an artistic enterprise that neither could have achieved working alone. We found such teams in all spheres of human endeavors: science, invention, exploration, athletics, art, architecture, entertainment, forensics – both factual and fictional – and many more.

Our focus narrowed to very small teams from the past 100 to 200 years, specifically those whose joint efforts changed the world. We wanted to know what made them tick.

Why did some accomplish a lot and others less so, or for a shorter period of time? Why did some spectacularly uncouple, like Gilbert and Sullivan, yet others stayed connected throughout their lives, like Susan B. Anthony and Elizabeth Stanton? The latter pairing, despite enormous personal differences at times, worked harmoniously for over fifty years. And others like the Arctic explorers, Admiral Peary and Matthew Henson, who persevered through innumerable polar hardships, but parted ways soon after.

A shared passion for a particular goal does not ensure that the team members will even like each other. Nor does it mean that the team will continue after the goal has been reached. Some teams remain together refining their work, but others part ways professionally to pursue separate interests, while maintaining a collegial relationship over time. There is room for all these types of teams and more.

As we researched this topic, we came to the realization that there was much to learn from these very small teams that could be useful in our own lives. How can we all become better team players and team leaders? How can we form teams that consistently exceed expectations?

Our general thinking became organized around a few probing questions:

1. What was the very small team's mission?

Did the team have a focal point? Were they trying to accomplish something that had never been attempted before? Was there danger – to the team members, or perhaps to the world at large – if something went wrong?

We studied where and how our subjects fitted into the world at the height of their success, and asked if what they did changed the world for the better.

Did the team's mission result in success as planned, or failure, or did they arrive at an entirely different outcome from the one first sought?

2. What were the origins of the team?

One of the areas we were curious about was whether the mission was chosen by the team, or if it was chosen for them by someone else who then put the team together. Did the team members have similar backgrounds and experience, or diverse upbringings?

We wanted to know if the team members saw themselves as equals, or if one was superior in leadership or stature. Some teams could have had a mentor-mentee relationship and we wondered how this had played out over time. A few teams we studied had one member who was a genius. How did the other team members adjust?

In some of the pairings the participants were involved in an intimate relationship. Would husband-and-wife teams like Captain Josiah and Eleanor Creesy work in entirely different ways from non-spousal teams? Would a sibling pairing like the Wright brothers work in a different way from a father-son team, like that of George and Robert Stephenson, the creators of locomotives and railways?

3. What was the team's durability?

Would the team withstand the test of time? Was their pairing a one-mission association like some of the Apollo space teams, or long-term over many missions? Was success a determinant of the team's durability and longevity? Did it lead to further missions by the team, or did success resulting in wealth become a source of friction?

We were interested in the harmony or strife within the working relationship. Did team members carry their experiences with one team into other team-based enterprises? Did their work end in a happy, rewarding manner, or otherwise? And even if the partnership dissolved, did they remain close afterwards?

And finally, what was the very small team's legacy for the world? Why and how are they remembered today?

Let's get started

We started with so many questions, and we hope you have your own that has spurred further interest in the teams we will discuss.

There is something for everyone in this book, and while we focus on ten core teams who worked in a variety of exciting endeavors, there are many more teams mentioned or discussed. The last two chapters in the book provide insights about what we can learn from all of these amazing very small teams. To aid your appreciation of the time spans that these teams were working in and their degree of overlap with one another, the book's appendix contains a timeline.

Now, let's dive right in.

Where better to start than looking at one very small team: Susan B. Anthony and Elizabeth Cady Stanton. In the 1800s, they bravely championed women's rights at a time when women were barely considered to be citizens.

Part 1
Solving A Specific Problem

★

DECLARATION OF RIGHTS

OF THE

WOMEN OF THE UNITED STATES

BY THE

NATIONAL WOMAN SUFFRAGE ASSOCIATION,

JULY 4th, 1876.

WHILE the Nation is buoyant with patriotism, and all hearts are attuned to praise, it is with sorrow we come to strike the one discordant note, on this hundredth anniversary of our country's birth. When subjects of Kings, Emperors, and Czars, from the Old World, join in our National Jubilee, shall the women of the Republic refuse to lay their hands with benedictions on the nation's head? Surveying America's Exposition, surpassing in magnificence those of London, Paris, and Vienna, shall we not rejoice at the success of the youngest rival among the nations of the earth? May not our hearts, in unison with all, swell with pride at our great achievements as a people; our free speech, free press, free schools, free church, and the rapid progress we have made in material wealth, trade, commerce, and the inventive arts? And we do rejoice, in the success thus far, of our experiment of self-government. Our faith is firm and unwavering in the broad principles of human rights, proclaimed in 1776, not only as abstract truths, but as the corner stones of a republic. Yet, we cannot forget, even in this glad hour, that while all men of every race, and clime, and condition, have been invested with the full rights of citizenship, under our hospitable flag, all women still suffer the degradation of disfranchisement.

The history of our country the past hundred years, has been a series of assumptions and usurpations of power over woman, in direct opposition to the principles of just government, acknowledged by the United States at its foundation, which are:

First. The natural rights of each individual.

Second. The exact equality of these rights.

Third. That these rights, when not delegated by the individual, are retained by the individual.

Fourth. That no person can exercise the rights of others without delegated authority.

Fifth. That the non-use of these rights does not destroy them.

And for the violation of these fundamental principles of our Government, we arraign our rulers on this 4th day of July, 1876,—and these are our

ARTICLES OF IMPEACHMENT.

BILLS OF ATTAINDER have been passed by the introduction of the word "male" into all the State constitutions, denying to woman the right of suffrage, and thereby making sex a crime—an exercise of power clearly forbidden in Article 1st, Sections 9th and 10th of the United States Constitution.

The Women's Declaration of Rights, written on 100th anniversary of the founding of the United States. *(Library of Congress)*

Chapter 1
Women's Rights

Susan B. Anthony and Elizabeth Cady Stanton

"The days when the struggle was the hardest and the fight
the thickest; when the whole world was against us, and we
had to stand the closer to each other."

—*Susan B. Anthony, 1902*

O f all the very small teams we studied for this book, we found
none more awe-inspiring than Susan B. Anthony and Elizabeth
Cady Stanton.

The transformational power of the Wright brothers' airplane,
the adventures of world's bravest explorers, the cultural impact
of Gilbert and Sullivan – these almost pale in significance to the
changes that Susan and Elizabeth brought to the world. Especially
when you consider the lack of women's rights when they first met,
and how these rights had changed forever when they said goodbye,
51 years later.

In this chapter we have taken the liberty of referring to the two
women by their first names, not to diminish their stature but to
enhance it – to remind our readers that they were women. To refer
to Susan B. Anthony by only her last name, Anthony, and to refer to
Elizabeth Cady Stanton as Stanton – both surnames being effectively

masculine in common usage as first names – is to deny the full impact of what Susan and Elizabeth achieved in an era when the social status of women was far below that of men. We put Susan's name first because she became the most famous of the pair, but as you will see, they were true equals.

In comparing their actions with those of other small teams in this book, we found striking contrasts and similarities. Also, because women's rights continue to be one of the most important societal and political issues in the United States and the world today, we have, at times, included references to modern events.

The starting point

Everything we know of as women's rights today was universally denied to women in the early 1800s. The whole idea was unheard of in that era.

The concepts that women should have control over their health, attain a university education, own money, property or bank accounts not under the control of their husband, or work outside the home did not exist when Susan and Elizabeth began their work. There were very few jobs for women, and those that were could only be found in well-defined professions such as nursing or teaching children. Any income from such work went directly to the husband, to own and manage.

The most basic of rights – to choose what clothing to wear, to ride a bicycle, to divorce an abusive husband, to speak in a public forum, to earn equal pay for equivalent work done by men – were denied, if not by law then by custom. The list goes on and on.

Women were not just second-class citizens. In the United States and the world, they were barely citizens at all.[1]

---

[1] Proof of this is the wording used in the US Declaration of Independence, dated July 4, 1776. The second use of the word 'Men' with the letter M capitalized was exactly how it was written in the original document: "We hold these truths to be self-evident, that all men are created equal, that they are endowed by their Creator with certain unalienable Rights, that among these are Life, Liberty and the pursuit of Happiness. That to secure these rights, Governments are instituted among Men, deriving their just powers from the consent of the governed."

Every law was made by men

Women did not have the right to vote. Every law was made in legislatures comprised entirely of men. Regulations were approved by men and voted on by men. These became enshrined in common law – even those that directly affected a woman's well-being and livelihood.

The only way for women to change this, including the right to vote and access to all basic justices, required the explicit approval or vote of only men, specifically white men.

In the United States for example, white men controlled the presidency, governorships, legislatures, and courthouses. The constitutions of the states, and of the federal government, were all written to ensure that all second-class citizens – that is, everyone who was not born a white male, were denied basic rights – including the right to vote.

Other societal ills existed when Susan and Elizabeth began their fight in 1851. The most significant was the existence of slavery in the United States. Feelings on both sides of the issue were so strong that they would lead to the potential break-up of the United States. The outcome produced a Civil War and the death of over 600,000 Americans.

Another was rampant alcoholism among men, some of whom were prone to drinking away their weekly wage. Wives in those situations had no say in how their husbands spent their money, had no legal recourse, could not divorce, and had no private funds to leave their husband. They had no rights to raise their own children if they chose to leave. Wives had no defense in the law and those who were abused in their family home had no societal protections whatsoever.

In fighting for women's rights, Susan and Elizabeth were at times leaders in both the anti-slavery movement and the temperance (drinking in moderation or alcohol abstinence) movement, but they knew absolutely that the key to everything would be the right to vote. It would be through an exercise called suffrage that would enable all of society to elect better candidates – including women –

with the aim of bringing about an overwhelmingly positive change in society.

But there was another challenge they were not prepared for.

Not only did powerful men of that era not want to give women rights, many women also did not want those rights. They did not want equal pay, equal status in a marriage or even the right to vote. Many feared that these elements would lead to an imbalance in society and weaken their role within it.

Shared passions, individual lives

When Susan and Elizabeth first met in 1851, they soon realized that they shared a common passion for promoting the rights of women in the political arena. Each already had her own well-developed viewpoint and passion in the matter of women's suffrage and was not afraid to express herself publicly. But two voices speaking together are louder, and will always get more attention than those same voices speaking alone.

As individuals in a team, Susan and Elizabeth were different: Elizabeth was married with seven children. Susan had offers of marriage from suitors but turned all of them down. The women lived apart but were inseparable in their common goals. They didn't always see eye-to-eye. As you will soon learn, they had strong disagreements on key topics, arguments that they hid from others in an effort to protect their valued friendship and public position.

Sometimes, when one would lose faith in the project, the other would bolster. They weren't facing imminent disaster like some other very small teams you will meet later in the book, but some of the numerous and frequent attacks and setbacks they endured were almost as daunting. And just like every duo and trio in this book, they faced each obstacle with determination, raising their game and coming back stronger.

In an era before telephones and electricity in the home, the only means of long-distance communication was by written letter, taking days or weeks to reach its destination. Their thriving and voluminous correspondence over decades has left us with hundreds of letters that give deep insight into their friendship, their travels together and apart, and their struggles.

Conventions

Susan and Elizabeth began their extensive career by traveling around the US. They set up conventions promoting the importance of women's rights more than a decade before the railroad network was developed. It was only later that the railroads would link major cities in the east and midwest with the remote and barely-settled territories of Wyoming and Utah, and other locations still aspiring to statehood farther west. In the absence of a transcontinental railroad, they endured weather extremes as low as -20° F (-29° C). They were doing this in the cloth and woolen clothing of the time, and they continued to do these types of journeys right into their advanced years, when both women were over 70 years old, and during some of Susan's later travels, when she was over 80 years old.

We write about small teams of brave explorers and inventors risking their lives: Peary and Henson in the Arctic, Shackleton and Wild in the Antarctic, Hillary and Tenzing on Everest, and Wilbur and Orville Wright on every flight they took. Susan and Elizabeth were also fearless in the face of resolute and dangerous opposition to their liberal ideas. They set up conventions and gave talks, sometimes to armed and openly hostile audiences, risking their lives to get their message heard. After some tumultuous situations, at one talk, the mayor of Albany in New York was seated on stage while Susan gave a talk, holding a gun to ensure her safety.

Fifty years later, this national convention, renamed the National American Women Suffrage Association, was still going strong, as were Susan and Elizabeth. They never tired. They never gave up the fight. Remarkably, Susan or Elizabeth served separately or together as leaders and keynote speakers at almost every one of these fifty conventions.

Teamwork

The surviving photographs of the two women show their facial

Elizabeth Cady Stanton, seated, and Susan B. Anthony, standing on the right.
(Library of Congress)

expressions to be rather dour, but this was as a result of the long exposure time needed for portrait photography in those days. They were in fact energetic women full of vigor and with a passion for life, and were gripped by the importance of their work and collaboration.

Susan and Elizabeth had tremendous written and oratory skills, honed over the years, and were masters at explaining their feelings in their letters, as well as the importance of being able to lean on each other. Men in that era, even the most expressive of them, seldom wrote so intimately.

They wrote often about their friendship, their teamwork, the division of duties, and their respect and love for one another, leaving a lively record of individually authored and co-authored speeches, articles and books. As friends, they also frequently corresponded with each other.

Another aspect of their teamwork confounded our desire to

pigeonhole them. Among the other teams in our book there were clear leaders: *Sir* Ernest Shackleton, *Admiral* Robert Peary, *Captain* Josiah Creesy, *Commander* Jim Lovell, each with the designation and credentials to claim it. Even among the equal Wright brothers, Wilbur was the genius of the two. Susan B. Anthony and Elizabeth Cady Stanton were equals. Neither was the leader.

They were equally talented speech writers and orators. In the early days, when Elizabeth's many young children needed her attention, she could write speeches but not travel to conventions. Susan had to go alone and set them up, and then address the audience herself. Their roles became defined during that period, but whether the speech was written by one, the other, or both, it had the power to hold the rapt attention of an audience.

They were true pioneers in every sense of the word, climbing the tallest 'societal' mountain, but unlike Hillary and Tenzing at Mt. Everest, Susan and Elizabeth did not have 30 years of detailed climbs to learn from or aerial views of the 'terrain,' or a support system of hundreds of people specifically there to aid them.

And on every step of Susan and Elizabeth's long lives driving the women's rights movement, whenever they encountered naysayers and critics and others who exclaimed, "This can't be done," they not only thwarted them, at times, they won them over to champion their cause.

The journey

Whether you are flying to the moon, sailing a clipper ship around the bottom of South America, climbing Everest, or striving to get to the South or North Pole, it's imperative to enjoy the journey. As Hillary and Tenzing found out, you are only on the summit for 18 minutes at best – a mere fraction of the time it took to get there.

The same was true for the women's rights movement. Securing the right to vote was a long and arduous journey, and participants, agitators and leaders like Susan and Elizabeth focused on enjoying the thrill of the journey. Celebrations at the top were few, far between, and at times fleeting.

Chapter 2
Women's Rights

Dynamic From The Start

Although the term "feminist" had already been coined by 1852, it did not come into general usage until the second feminist wave swept through in the 1960s. Elizabeth and Susan were, by all accounts, the leading feminists of their day. Born in November 1815, Elizabeth from an early age showed herself to be just that. When she married at the age of 25, she insisted that her marriage vows to Henry Stanton be rewritten to omit the word 'obey.' She wanted a right that was unheard of in 1840 – to be a married woman and not be subservient to a husband.

Elizabeth was socially and politically active, and became known for her outspoken, direct articles about women's rights for a temperance newspaper called *The Lily.*[2] She had three young children in the house by the time she was 28 and living in Seneca Falls, New York.

Elizabeth's domestic duties didn't stop her from joining forces with Lucretia Mott to create a two-day gathering for women's rights in July 1848, called the Seneca Falls Convention. This was the first such convention – we'd use the word conference today – designed especially for women, with sessions on law and the role of women in society. It was followed soon after by similar conventions in the coming years, up until the outbreak of the Civil War.

---

[2] *The Lily*, the first newspaper for women, was issued from 1849 until 1853.

In a telling demonstration of her writing prowess, feistiness, and determination, Elizabeth used the Convention's platform to debut her *Declaration of Sentiments,* loosely based in structure on the United States' 1776 *Declaration of Independence.* The original had no mention of women; Elizabeth's version expanded on the original text, with the words 'all men *and women* are created equal,' along with a demand: women must be granted the right to vote.

Her husband said he'd leave town if she presented it. She did. He left. They overcame that difficulty. Their marriage was long and bore seven children. Despite their partnership, they clearly had differences of opinion, and Elizabeth was not afraid to voice hers. Her declaration was signed by 100 attendees, over 30 of whom were men.

The event was so successful that another women's rights event was organized in Rochester, New York. Susan B. Anthony did not attend that convention, but her parents and sister did. At the time, Susan was building her own social activism credentials by giving speeches about the need for temperance.

Two years later in 1850, 30-year-old Elizabeth was invited to participate in the first Ohio Women's Convention, but by this time she had multiple children to look after and could not attend.

Elizabeth had already begun to build a name for herself in women's rights and other issues, while 25-year-old Susan was becoming known for her stances on correcting social ills. Susan was as energetic about temperance as Elizabeth was about women's rights.

They were coming from different directions and arriving at a similar place.

Elizabeth submitted a keynote speech in the form of a letter to be read aloud to the 500 attendees at the Ohio Convention. This became her only way of connecting with the conventions she could not attend. Unable to attend the even bigger First National Women's Rights Convention in Worcester, Massachusetts six months later, she instead contributed a letter that was read to the 1,000 women and men in the audience.

Although the two women had activist friends in common – the editor of *The Lily* Amelia Bloomer, the anti-slavery women's rights activist Amy Post, and the Black former slave and anti-slavery campaigner Frederick Douglas – Susan and Elizabeth had yet to meet.

That all changed in May 1851 at an anti-slavery meeting in Seneca Falls, when Amelia Bloomer introduced them.

Partnership begins

Outraged that women were not allowed to speak (except to the women's auxiliary) at a meeting of the Sons of Temperance in 1850, Susan organized her own conference in 1852. She created the Women's New York State Temperance Society Convention, and deemed men could attend but not speak at the convention.

Susan invited Elizabeth to serve as the president of the organization. Five hundred women attended the event in Rochester, New York. While this convention was about reducing alcoholism and the effects of alcohol on society, Elizabeth dared to raise issues such as women being permitted to divorce abusive drunk husbands. Prior to this, with women being deemed subservient to men, they would have just had to endure such terrible treatment.

Many more conventions were happening at state level in Ohio, Indiana and Pennsylvania, and at national level in Worcester, Massachusetts and Syracuse, New York. Elizabeth had been writing speeches to be delivered by others for each conference, but at Syracuse, a new pattern emerged. Susan became the orator, reading Elizabeth's passionately written words and putting a voice to her radical ideas. Ideas like pressing for the transformation of university admissions to accept women.

How radical was this idea?

Princeton University, founded in 1746, didn't start accepting women until 1969, over 100 years *after* Elizabeth proposed it in a speech read by Susan. Even then, the idea was vehemently opposed by a group called the Concerned Alumni of Princeton. One hundred and fifty years after Elizabeth's speech, Princeton finally achieved a milestone, with a freshman class comprised of fifty percent women and fifty percent men.

Progress is hard. But it only starts when bold people like Susan and Elizabeth match political action, influencing social thought by putting their brave message out there, regardless of the threats that ensue. And this, as shown by them, is best done by a very small team.

The pattern they would follow, with Susan reading Elizabeth's words to a large audience while Elizabeth was home with a growing family, went on for many years at many conventions around the nation. Her young children needed her attention, but at the same time, Elizabeth was nurturing the boldest ideas for American women's rights yet to be voiced in public and putting them into compelling speeches that would hopefully drive action.

Susan, recognizing the power of their teamwork, had a challenge. She didn't always agree with everything that Elizabeth wrote, but as the orator of Elizabeth's words, she had to speak them as written. To gain forward momentum in their very small team meant give-and-take by each of them to achieve their common goal.

But going forward, they needed to devise a different strategy. Susan agreed to visit Elizabeth at her home, where they could discuss their ideas together. Susan could look after Elizabeth's children while Elizabeth wrote the speeches.

The first setback

While both were focused on women's rights, Susan also kept her focus on reducing societal alcohol consumption. Since no existing temperance society would allow women to be members, in 1852, she and Elizabeth formed their own. The Women's New York State Temperance Society would go on to petition the New York State legislature to pass a law limiting the sale of liquor. Their petition was rejected out of hand because most of the 28,000 signatures belonged to women and children.

At the society's first annual convention in 1853, it was agreed to allow men to have leadership roles. Since men would now be influencing decision-making, the word "Women" in the Society's name was changed to "People". The focus on women's rights, which Susan and Elizabeth had worked so hard to achieve, was now in the dustbin.

This organization – set up and run by women, yet open-minded enough to allow men on their board – watched helplessly as the men took over. They drowned out the women's voices, dismissed the women's ideas, renamed the organization, and declared it their own. Modern women will recognize this as a story as old as time.

The impact of what happened hit Susan harder than Elizabeth.

In some of the teams in this book, when one team member becomes discouraged, the other lifts them out of despondency. When Wilbur Wright lamented after one of the Wright brothers' glider failures, Orville boosted his spirits and belief in the project. After the takeover of the Women's State Temperance Society, Elizabeth proved the resilient one, rallying Susan to regain focus on the goal.

Focusing on women's rights

Susan began focusing more on women's rights than temperance, and on changing the law so it would grant women equal pay for equal work, and assign property rights similar to those given to men.

In 1854, Susan planned a convention in New York State where she and Elizabeth both lived, with the intention of identifying the laws that were worst for women and changing them. She enlisted Elizabeth who had greater legal knowledge than her to create a set of demands they could take to the state legislature. It would call for women to have equal civil status as men. The common right to vote and sit on juries, or to own real estate and control personal bank accounts did not exist for women in 1854.

Although preoccupied with family priorities, Elizabeth was more than equal to the task. She travelled to the state capital in Albany and delivered one of the most important speeches ever given there, calling on the legislature to open the way to change. She made her key points one by one, enumerating the existing injustices, and proposing solutions: "How, I ask you, can that be called justice, which makes such a distinction as this between man and woman?"

she asked the all-male legislature.

To help instill their desired change, Susan printed and distributed 50,000 copies of the speech.

It made no difference. Nothing changed.

Two years later at a teachers' convention, Elizabeth's speech, co-written with Susan, presented the case for boys and girls of the same age to receive the same education. They raised the issue that women teachers should have equal pay to male teachers only to be soundly rebuffed by the women teachers, who felt it would change the balance in society. The very people who would benefit most disagreed with their ideas and recommendations.

Again, setbacks hurt, but they continued on.

Although it seems almost unbelievable today that equal pay rights would have been so opposed by *any* percentage of women, we are seeing a similar scenario playing out with abortion choice and health care rights for women. These are being cancelled or strongly curtailed in some US states where a percentage of American women are willing to give these rights up.

Their task was not unlike that of Hillary, Tenzing and the team of climbers and Sherpas moving supplies up the mountain of Everest to reach even higher altitude camps. The support teams had to go back down the mountain, recover from the work and stress of the task they had set themselves, then do it all again a few days later.

In ascending their 'mountain' to achieve women's rights, Susan and Elizabeth had to navigate treacherous icefalls and untried routes, only making it partway up before having to retreat, regroup, and re-plan a new route.

At times, their uphill struggles must have felt endless.

But to reach the goal – the mountain top – they knew, just like all the other very small teams in this book did, it would be a long and arduous journey. Only perseverance, resilience and optimism would get them through.

Chapter 3
Women's Rights

Stamina

With Elizabeth still burdened with family responsibilities in 1857, and the anti-slavery movement gaining strength in the lead up to the Civil War, Elizabeth stayed home while Susan gave anti-slavery lectures across the country. The lecture tours were grueling – arduous traveling; hours of speaking; staying in poor accommodation with even poorer food. Her schedule was so harsh, other speakers and organizers on the tour dropped out when the work became too much. But not Susan.

Three years later, their perseverance was starting to pay off. In 1860, the New York State legislature passed the Married Women Property Act enabling women to manage their own business, property and money. But despite the sweet taste of victory, Susan and Elizabeth wanted more.

They wanted women to have the right to vote.

It was only through the ballot box that social reformers could be elected, reformers who could benefit society, women, and families and eliminate slavery, alcoholism, and poverty – the big social ills of the day.

In the years since 1850 when the First National Women's Rights Convention was held in Worcester, Massachusetts, there had been nine of these annual conferences. Each was aptly named the Second National Women's Rights Convention, the Third National Women's Rights Convention, and so on. Susan, Elizabeth and their support staff were gearing up for the tenth.

With her children more grown up, Elizabeth could finally attend. She gave a passionate speech, in which she demanded that every woman should have the right to divorce an abusive husband.

Eventually there would be fifty of these annual conventions, consecutively numbered, and either or both Elizabeth and Susan would be at almost all of them. Although they had not yet reached the summit, they were making progress.

Risk and life-threatening danger

The following year, Susan used her organizational talents to assemble a small team of dynamic lecturers who would do talks on the topics of the day: abolishing slavery, freeing slaves, and women's rights. Elizabeth was the speaker on woman's rights. Their target was audiences across New York State.

The pair had both lived in New York State for a long time. Their reputations had grown to the point that they were seen as agitators. To those who believed what they were advocating for, they were heroes. To the men and women who were still vehemently opposed to their ideas, Susan and Elizabeth were the agents of change, chaos and an uncertain future. This placed the pair in mortal danger.

Animosity grew. Audiences were hostile. Figures of them were burned in effigy.

Dangers increased as audience members could be armed, angry and easily incited to violence. An event in the state's capital city Albany was peaceful only because the mayor ensured police were visible to the audience while he sat on the stage with a loaded firearm.

A four-year disagreement

Civil War in the United States became unavoidable, caused by a clear divide between those who wanted slavery (white people located in southern US states) and those who found it abhorrent (the vast majority of those living in northern states). Abraham Lincoln, a northerner, won the presidency in November 1860, and the southern states began to secede from the union.

By the end of January 1861, Alabama, Florida, Georgia, Louisiana, Mississippi, South Carolina, and Texas had left, and organized their own national government called the Confederate States of America. In an act of open warfare against their northern neighbors, the Confederates attacked Fort Sumter in Charlestown, South Carolina. Arkansas, North Carolina, Tennessee, and Virginia soon joined the confederacy, and the four-year Civil War began in earnest.

A smaller rift, waged without weapons, began to open between Susan and Elizabeth. Elizabeth wanted to focus their work on the anti-slavery message and put aside their quest for women's rights for the duration of the war, however long that might be. She felt that when the North won the war and slavery was abolished, they'd have proven to be on the correct side. The powerful forces opposing the recognition of women's rights would be grateful and grant women the rights they had so long been denied, including the right to vote.

Susan strenuously disagreed.

Although Elizabeth was older and had more real-world experience than Susan, she had mostly been fighting for women's rights from her desk, in the form of letters, speeches and articles. Susan had been on the front lines, chairing the conventions, giving the keynote speeches, traveling from venue to venue, and was on the receiving end of widespread vitriol from men and women who didn't like her message or Elizabeth's words.

In Susan's view, Elizabeth's plan to sit on the sidelines for the duration of the war was time wasted that could be spent championing women's rights.

There would be more disagreements between them, but so far, this was the most fundamental and the most serious. A lesser team would have broken up, like Gilbert and Sullivan who came apart over a petty issue after ten years working together. But in the 1860s, Susan and Elizabeth knew their cause and their relationship were incredibly valuable. They agreed to put aside their differences. One or the other would eventually be proven wrong.

In the meantime, they had work to do. And it was best done together.

Susan moved in with Elizabeth and her family in New York City. It was during this time that they formed the Women's Loyal National League to ensure that the end of the war would end slavery in the United States forever by creating a constitutional amendment. The pair used the women in the league to collect over half a million signatures on a petition that was given to Congress. Again, their efforts failed.

By the end of war, Elizabeth came to realize the seriousness of the error in her beliefs. Susan had been right. Men in power did not feel they owed women anything. Nothing changed.

They were right back where they were at the start of the war – working as hard as ever to rally support for women's rights.

Undeterred

One year after the end of the Civil War, the pair created a new conference, dubbed the Eleventh National Women's Rights Convention. They had a new cause to fight for: universal voting rights not just for women, but also for Black men and women.

Even though women could not vote, there was nothing stopping them from running for elected office. Elizabeth ran for Congress in 1866 but did not win. (The first woman to win a congressional seat occurred more than 50 years later.) Also in 1866, slavery was abolished by the passage of the 14th amendment to the United States Constitution.

Susan and Elizabeth now were back to fighting state by state for women's right to vote. They campaigned in Kansas, which was not yet a state. As with all elections, only white men could vote. There were two resolutions: one to grant Black men the right to vote, the other to grant women the right to vote. Both were defeated. Vote counts revealed more white men were willing to give Black men the right to vote, than were willing to give women (white or Black) the right to vote.

Susan and Elizabeth were fighting an uphill battle. Black men and former slaves were given the right to vote in the 15th amendment, which was passed in 1870 – the same year that saw the twentieth anniversary of the First National Women's Convention in Worcester, Massachusetts. To add insult to their efforts, which included appeals to Congress, they were again reminded that women could not have the right to vote because they were merely members of states. They were not actual citizens.

Still, Susan and Elizabeth did not give up.

They established a National Woman Suffrage Convention a few years later and formed the National Women Suffrage Association.

A new tactic

In 1871, Susan and Elizabeth and other suffrage supporters tried a new tactic. The wording of the 14th and 15th amendments meant they could be considered together to grant women the right to vote because the two bills granted freed slaves the right to vote without restricting this to just freed male slaves. A Congressional committee disagreed. Susan and Elizabeth encouraged women to try to vote anyway.

The pair had a few more rifts. When a suffragette named Victoria Woodhull ran for President of the United States in 1872 – women still did not have the right to vote – Elizabeth supported Victoria's right to campaign.

Susan opposed it.

The Republican Party added a statement to their platform saying that women's demand for equal rights should be given 'respectful consideration.' Susan welcomed this despite the weak phrasing.

Elizabeth opposed it.

In 1873, true to her nature and risking imprisonment, Susan tried to vote. She was arrested and fined $100, equivalent to over $2,500 today.

Progress

B oth women were becoming enormously influential. Through their efforts, women in general had achieved rights to work in a wide variety of jobs, attend colleges and universities, own property, earn money that didn't have to be given to their husbands, divorce abusive husbands, and have more control over what they wore.

Some states and territories were even allowing women to vote in certain elections, and in the territories of Wyoming and Utah women could vote in all elections. But Susan and Elizabeth still had not achieved the Holy Grail – women as equals to men, and women having the universal right to vote.

There were constant reminders of how far they still needed to go. For example, women were not included as speakers in the 1876 centennial celebrations in Philadelphia marking 100 years since the signing of the Declaration of Independence. Its famous words included:

> "We hold these truths to be self-evident, that all men are created equal, that they are endowed by their Creator with certain unalienable Rights, that among these are Life, Liberty and the pursuit of Happiness."

But this omission only made the pioneers even more determined. Teaming up with several women, they wrote a Women's Declaration of Rights and presented it to the Vice President of the United States.

But for every step forward, there was a step back. The Republican and Democrat platforms in 1880 for the presidential race did not mention women's rights. But in 1882, there was a step forward when a Senate Select Committee was formed to focus on whether women should be given the right to vote. In another victory the governor of the Washington territory (the area that now forms the state of Washington) granted women the right to vote and hold political positions in government.

In 1886 and 1887, men in Oregon and Rhode Island voted on referendums to give women the right to vote. Both referendums were defeated.

Then to add insult to injury, Washington and Utah withdrew the right to vote from women.

A new organization

In 1888 and with the stakes high, Susan and Elizabeth, together with others, formed the National American Woman Suffrage Association (NAWSA), which amalgamated the various factions of women's rights organizations already existing. At the age of 68, well above the life expectancy for women of that era, Elizabeth was made its first president. She and Susan had no intention of slowing down.

Another major success happened a few years later when the Wyoming territory was coming up for statehood.

There was a big question looming. Since women in the territory already had the right to vote in elections, would that be repealed when statehood was granted? President Harrison said it would not, and Wyoming in 1890 became the first US state that enabled women to vote.

At the same time, South Dakota was holding a referendum and despite Susan's tireless in-person campaigning there, the South Dakota men voted no.

One of the most powerful speeches ever

In 1892, at the age of 72, Elizabeth wrote one of the most powerful speeches ever given on American soil. The speech was titled *Solitude of Self* and was given both at the NAWSA convention and to the US

Senate judiciary committee.

In giving this remarkable speech, Susan was not demanding the right to vote. It was far more cerebral than that.

It told a story of the need for self-sufficiency of women that could be achieved through greater access to education, and greater recognition of the strength of women. It told the story of life, of women at various stages in their life, from girl to adult to becoming a widow, and then to old age. Elizabeth asked that women not be viewed as helpless individuals dependent upon men.

Here is a brief excerpt.

> "… No matter how much women prefer to lean, to be protected and supported, nor how much men desire to have them do so, they must make the voyage of life alone, and for safety in an emergency they must know something of the laws of navigation.
>
> To guide our own craft, we must be captain, pilot, engineer; with chart and compass to stand at the wheel; to watch the wind and waves and know when to take in the sail, and to read the signs in the firmament over all.
>
> It matters not whether the solitary voyager is man or woman. Nature having endowed them equally leaves them to their own skill and judgment in the hour of danger, and, if not equal to the occasion, alike they perish."

Her speech invoked another story in this book, that of Eleanor Creesy in 1851 as navigator of the *Flying Cloud* watching the wind and waves, confident in her ability to bring the ship to port. Whether Elizabeth knew of Eleanor Creesy's achievement is unknown.

Elizabeth's speech painted an evocative picture. Susan arranged for 10,000 copies of the speech to be printed and distributed, making sure that United States Senators each received a copy.

But as powerful as the speech was, it did not have an immediate impact.

Elizabeth, Susan and their team had long been working toward a 16th amendment that would grant women the right to vote. Their speech did not make that happen. Despite Elizabeth's impressive oratory, their efforts were still stymied by women who did not want the vote, and who did not want to be self-sufficient, educated, or independent of men. And these anti-suffragette voices were only getting louder.

Cartoon showing President Grover Cleveland, carrying a book titled, "What I know about women's clubs," being chased by Susan B. Anthony wielding an umbrella as Uncle Sam laughs in the background. *(Charles Lewis Bartholomew 1892-1896)* *(Library of Congress)*

The following year, 1893, Colorado men voted to give women the vote. Out of 44 states that comprised the USA at the time, only two permitted women to vote: Wyoming and Colorado.

New disagreement

In one of our previous books, *Audacious Goals, Remarkable Results*, we showed that people who performed at epic levels never retire. They don't stop to rest on their laurels or bask in the praise of achievement. They just keep on going to pursue their goals, no matter their age. Elizabeth and Susan, well into their 70s and 80s, were those people, and kept at their quest for women's rights.

But it wasn't all plain sailing. Sometimes, the pair vehemently disagreed.

Elizabeth believed that one of the things holding women back was how men interpreted the Bible. So she wrote a Women's Bible, which interpreted the stories from a woman's viewpoint. This was a monumental undertaking considering how strenuously Susan objected to it. Susan did not believe religion was to blame for the predicament of women rights.

Given Elizabeth's age, travel was becoming harder for her and Susan agreed to go in her place. She read Elizabeth's speeches to audiences even when she did not personally agree with the sentiment. But she did this without objection. Because their friendship mattered, and she believed that friends were entitled to have differing viewpoints on key topics and still be able to work toward a common goal.

Age was no barrier

In 1896, California and Idaho men voted on whether to give women in their states the vote. California said no. Idaho said yes. In 1897, the same year polar explorers Peary and Henson (a team discussed in this book) were battling Arctic weather, Susan at the age of 72 traveled to Iowa amid a raging blizzard to campaign for women's rights. In that same year the Utah territory, which had already allowed women to vote, became a state and kept that right intact.

1898 was the 50th anniversary of the First Women's Rights Convention in Seneca Falls, New York, but Elizabeth was now too old to attend. A few years later Susan, who had taken over the presidency of NAWSA, stepped down from that role.

Even after her dear friend and colleague Elizabeth passed away, Susan kept the women's rights torch burning well into her 80s traveling to NAWSA conventions in New Orleans and Washington DC. In 1904, she journeyed to Berlin, Germany to attend the International Council of Women.

In the following years she attended conventions in Portland, Oregon and Baltimore and passed away in 1906, three years after the remarkable, very small team of the Wright brothers invented the airplane, a device that many years later would have made Elizabeth's and Susan's travel to all the various conventions so much easier.

The Amendment

Susan and Elizabeth had hoped that the women's vote would be enshrined in the 16th amendment. It wasn't to be.

When the 16th amendment passed several years after Susan died, that amendment granted the government the right to impose taxes on citizens. The 17th amendment gave (male) voters the right to vote directly for Senators. The 18th amendment prohibited the sale of alcohol.

Finally, the 19th amendment, which granted women the long sought after right to vote, passed Congress in 1919. It was ratified in 1920.

An enduring legacy

The freedoms women have in the United States and even in certain countries around the world – from being a full citizen to working in any profession; being in the military; owning their own business, money, property and bank accounts; playing sports in public and riding bicycles; wearing what they want, all the way through to voting – stem from the tireless and courageous work of the very small team of Susan and Elizabeth.

While we write that these freedoms were won for women, Susan and Elizabeth's fight was also for women *and men* to have the right to move unhindered throughout society, to achieve all that can be possible in a free and democratic world.

Interpreted even more broadly – is that all people have freedom of the soul – to be your own person, able to navigate whatever waters you want to be in, and achieve all that you are capable of.

★

As Susan and Elizabeth were nearing the end of their quest for women's rights, two brothers were just getting started on theirs. They were attempting to invent a flying machine.

Orville Wright *(Left)* aged 34 years, Wibur Wright *(Right)* aged 38 years, about 1905.
(*Library of Congress*)

Chapter 5
Flying Machines

The Wright Brothers

"…nearly everything that was done in our lives has been the result of
conversations, suggestions and discussions between us."

—*Wilbur Wright*

Great inventions have often been the brainchild of one person working alone, but throughout history there have been notable exceptions.

Two people collaborating in business enterprises have led to highly successful brands bearing their names: Hewlett-Packard, Procter and Gamble, and Ben & Jerry's. Two co-founders working toward a common goal created the foundational inventions that turned Apple, Microsoft, Google and Intel into some of the most powerful and profitable corporations in the world today.[3]

Looking back another century we find other pioneering inventions by very small teams. The father-and-son team of George and Robert Stephenson built the first railway, the first passenger train and

---

[3] Steve Jobs and Steve Wozniak (Apple), Bill Gates and Paul Allen (Microsoft), Larry Page and Sergey Brin (Google), Gordon Moore and Bob Noyce (Intel).

designed the first locomotive to pull that train. Father-and-son Marc and Isambard Kingdom Brunel built the first tunnel under a flowing river. The tunnel, opened in the 1840s, not only changed urban life forever but their innovative construction technique is still used to this day. It has been used to build underwater bored tunnels like the Channel Tunnel joining Britain to France, and the Lincoln and Holland Tunnels connecting New Jersey to New York.

But no matter how great these stories of invention and cooperation were, or how important the changes brought to modern society, they pale in significance when compared to the next story.

This is the incredible tale of how a pair of brothers who owned a bicycle shop stunned the world by building and flying the first airplane. They did this at a time when governments, rich corporations and world-renowned scientists were unable to crack the mystery of flight. And all of the Wright brothers' work was self-funded on a very limited budget.

These two were uniquely talented, but there is so much more to the story. We shall reveal it when we examine how this high performing, very small team operated.

The challenge of manned flight

Before we even get to who Orville and Wilbur Wright were, let's take a look at the challenges of heavier-than-air manned flight.

Airplanes can't be made to fly by simply mimicking what a bird does. The dynamics of birds – their weight-to-lift ratio; weight-to-strength ratios; their wings, feathers, and lightweight skeletal structure – can't be replicated in a manufactured object.

Over the centuries people (mostly men, but some women, too) tried, building feathered, flapping wings from various lightweight materials, attaching them to their bodies, and jumping from high places, thinking they might fly. Some, tragically, plunged to their death.

For a flying machine to work it must be able to both float and move forward like a glider, with the power to launch using its own engine, to accelerate and decelerate while in flight, and to land without injuring the pilot, the passengers, or the aircraft itself.

Another challenge involves maneuverability.

A true flying machine has to be controllable left and right, up and down, and be able to twist and turn to reach any destination – not just in a straight-line linear path from the take-off point. It must be able to rise high enough to clear mountain ranges, safely descend to the ground wherever the pilot determines, and achieve a perfect landing on it *every single time*, even during inclement weather.

A desire over the centuries

Some of the greatest minds in history, including Leonardo da Vinci and Galileo, were intrigued by the challenges of flight and the complexity of aerodynamics. Over the centuries, aspiring flyers, scientists, crackpots, kooks, and daredevils experimented with kites, hot-air balloons, and wood-framed gliders. To learn more they studied birds in flight, and some like da Vinci even dissected birds and bats to understand their internal structure.

For all of Leonardo's inventive genius to envision what flying could be, the benefits it would accrue to governments and fighting armies, and even to design flying machines (he made over 500 sketches of flying machines or birds in flight), it is not clear that any of his designs were certain to work.

Mimicking what he saw in flying birds, da Vinci focused on building flapping wings to achieve flight, but without understanding that no human could sustain such motion to achieve lift off or control.

But Leonardo was on the right path. At a time when electricity in houses was still four hundred years in the future, he was designing helicopters and flying machines in his notebooks. He understood the importance of wing curvature to achieve lift for take-off and ascent, and that aerial control was dependent on the pilot's location and the center of gravity of the machine.

Despite all his effort, the mystery of flight remained just that – a mystery.

Balloons, toys, and gliders

Wilbur and Orville Wright were not the first brothers, nor the first very small team to venture into the world of flight. In the 1780s, the first hot-air balloon was built in France and sent aloft by another very small team. That team was brothers Joseph and Jacques Montgolfier.

Rather than putting themselves at risk, they put three farm animals – a duck, a rooster, and a sheep – into a basket attached to a balloon. This unusual assortment of animals set the first altitude and distance record, achieving a flying height of 6,000 feet (1.8 km), and covering a distance of one mile (1.6 km).

Next was a *manned* hot-air balloon flight. But the Montgolfier brothers didn't fly in it. They built the balloon for it, and then wisely turned the task over to two brave men, Jean-Francois Pilâtre de Rozier and Francois Laurent who amazed the citizens of Paris as they floated above them – airborne yes, but at the mercy of shifting winds.

In the late 1700s and early 1800s, British engineer George Cayley, who had been fascinated by flight from an early age, focused on heavier-than-air gliders. His prototypes pioneered many important discoveries required for lift and guidance: fixed and multiple-wing designs, and rudders and stabilizers for control.

Early in his career, Cayley built a children's toy in 1796 that appeared to fly or float in the air. It was based on a centuries-old Chinese top comprising of a stick shaft, topped with two fixed and rigid bamboo 'wings'. When the user spun the stick between their hands and let go, the top floated up toward the sky.

Cayley replaced the bamboo with wings of cork and feathers, added more wings top and bottom, and a strip of whalebone to start it spinning. This improved helicopter top could reach three times the height of the Chinese toy, achieving heights of 90 feet (27 m). Cayley built on this basic concept with new sketches, envisioning man-carrying gliders and a machine with four disc-shaped lifting devices like helicopter rotors (not unlike today's drones), capable of lifting a man into the air. Cayley was the first to realize that his glider could become a powered flight if a lightweight engine could be attached. He was significantly ahead of his time. Engine design at the time was still too primitive.

But his intriguing toy showed all who saw it or played with it that heavier-than-air flight was possible.

Cayley's design was later improved in 1870 by Alphonse Penaud, a Frenchman with a passion for the concept of flying machines. His helicopter top was made from paper and wood, and its spin was achieved with the twist of a rubber band. We mention this here because Penaud's toy becomes important later in this story.

In 1804, Cayley's first full-size glider was flown like a kite, but a manned flight did not occur until almost fifty years later. When Cayley was 80 years old, he commissioned a young man who worked for him to fly in one of Cayley's gliders, making that the first manned glider flight.

The dream of flight takes shape

By the 1870s and inspired by Cayley's published reports, other nations became interested in flight.

A German engineer, Otto Lilienthal, studied birds and flight, became an expert on wing design and air movement, and wrote a book about aerodynamics with detailed mathematical tables explaining lift and drag on objects like wings. These tables also become important later in this story.

In the 1890s, Lilienthal built many gliders of varying designs and flew them personally. It was becoming clear that to be successful in the field of flying, a person needed to be scientifically minded, detail oriented, exceptionally adept at building things, willing to accept that failure was part of the process, and be enormously brave. A crash could mean instant death.

His glider designs had dual-stacked wings and a vertical rudder for steering. Lilienthal was strapped to the glider in a way that enabled the lower half of his body to hang free below the wings. In a precarious position like that, each flight was seriously risky and after over 2,000 flights, his luck ran out. In 1896, he was gravely injured in a glider accident.

He died the following day.

Before his death, many of Lilienthal's glider flight experiments had been captured in photographic images, which further excited people about the possibilities of flying.

World leading scientists and inventors were also interested

Penaud's helicopter top had sparked the interest of many talented people. Nations and governments, particularly the French, were intrigued by the idea of flight. American inventors Thomas Edison (inventor of the phonograph and the incandescent light bulb) and Alexander Graham Bell (inventor of the telephone) were fascinated by the challenge, as was American-British inventor Hiram Maxim (who built the first automatic machine gun). French-born American civil engineer Octave Chanute, known for having built railways and the first bridge over the Missouri River, took up the challenge, as well as the astronomer and renowned builder of telescopes Samuel Langley.

All these men had proven track records, significant influence, personal wealth, and connections to the right people who could attract the funding needed to purchase raw materials and scientific equipment, hire construction crews and staff, and rent locations to carry out test flights.

In 1898, Langley was granted $50,000 by President McKinley (which equates to millions of dollars in today's currency) to expand his research and conduct test flights of his "aerodrome," a poorly performing, steam-powered flying machine.

Octave Chanute was the leading aviation inventor at that time. Edison ultimately had 1,093 patents to his name. All of them applied their creative minds and prodigious skills to build test models as well as real aircraft. They experimented using different wing designs, steam powered engines, and tried to achieve lift off by launching their glider from moving ships and railroad cars.

Each genius was making progress, but so far, none of them accomplished developing a successfully powered, heavier-than-air flying machine.

As they experimented, they discovered through trial and error that flying is far more than flinging an object into the air in a given direction. And it is significantly more than the drifting of lighter-than-air balloons above the streets of Paris.[4]

Launching a winged object into the air, even if it has the aerodynamic shape of a glider, is only a part of the overall program of flight. While all these men made progressive improvements, none were able to solve the fundamental problem of consistent take-off, controlled flight, and a guaranteed safe landing. But that did not stop them trying.

In 1899, into this mix of the greatest scientists and inventors of their day stepped two brothers. Their only claim to fame were being the proprietors of a small bicycle shop with Wright Cycle Co. printed in letters over the door.

---

[4] While the balloon's fabric, rope lines, and baskets give the object weight, the heated air inside the balloon is much lighter than the ambient air around it. This hot air provides the lift, so balloons are considered "lighter than air."

Chapter 6
Flying Machines

Passion and Perseverance

Wilbur was the eldest. As adults, they were inseparable. Neither married and they lived with their sister, Katherine. They ate together, ran the bicycle shop together and spent their leisure time together. Neither of them went to university, though Wilbur had been accepted to Yale University.

Before he could start at Yale though, an ice hockey incident resulted in Wilbur being smashed in the face with a hockey stick, knocking out many of his front teeth. It turned him into a recluse for several years and he never attended Yale. He spent his days instead reading books and caring for their ailing mother, who was dying of tuberculosis.

During Wilbur's reclusive years, Orville started a printing business and encouraged his brother to join him. They turned out a local newspaper, and then opportunity struck. When societal interest in bicycles began in earnest, the brothers opened a shop, selling, repairing, and ultimately designing bicycles for the many eager new customers taking up the sport.

As children, Wilbur and Orville had been intrigued by the Penaud toy helicopter their father had brought home from his travels as an itinerant clergyman. They had tried building larger models of it but with little success. As avid readers, they had a special interest in the successes and failures of aviation pioneers like Lilienthal, Langley and Chanute.

Both were familiar with Otto Lilienthal's experiments and had read articles and seen photographs showing him in one of his gliders. Lilienthal's death in 1896 lit a fuse in Wilbur, who developed an intense desire to understand how birds flew and how airplanes could be developed.

The two brothers were exceptionally bright, developing key skills and absorbing from their reading the complex descriptions of bird anatomy and wing structure. They spent hours standing together watching birds fly, discussing wing movements, and observing how birds like to fly in windy conditions. They learned to mimic a wide variety of bird movements with their arms.

In time they became two of the most knowledgeable people on the planet about flight, corresponding with the Smithsonian, Octave Chanute and others in the field, but none of these things made Wilbur and Orville unique in the brave new world of attempting to build an airplane.

What they did possess was a deep, innate curiosity driven by their upbringing, in a house where books, learning and craftsmanship were revered far more than wealth. Their father was a bishop in a local church. Their mother, who died when Wilbur was just twenty-two years old, was very skilled with tools and could make and repair everything, from toys to household appliances.

The brothers, both serious people, didn't seek the limelight, despite having exceptional written and spoken communication skills. Wilbur was tall, athletic, and good looking. Orville was smaller in height with a better business acumen, as evidenced by his starting of the printing business. Orville had even dropped out of high school.

This pair, measured against the greatest inventors of their generation, might be considered woefully under-educated. One was a reclusive high school graduate; the other a high school dropout. Both were from a lower to middle-class household living in a home that had an outhouse and no indoor plumbing.

But they had something none of the others had. They were a team. An exceptionally strong one. Their skills were complementary, but more important than that was their intense bond, borne out of a deep respect for each other's intellect and capability. Mix that with a passion for solving the challenge of flight while running their bicycle shop, and dogged perseverance to see all of this through, regardless of the risks.

The advantages of building bicycles

What also set Wilbur and Orville apart from Edison, Bell, Chanute, and Langley was that the brothers built bicycles. Not phonographs or telephones, or designed bridges or telescopes. The Wrights had hands-on skills in building objects that carried a human being, and literally moved that person from one location to another. One that was built from raw materials – metal for the frame, wood for the handlebars, leather for the seat.

They had the equipment and skills needed for measuring, cutting, grinding, shaping, fastening, and welding the raw materials into pieces that could be attached to one another. Their day-to-day work was assembling bicycles, followed by designing their own brand of bicycles and manufacturing them. The Wright brothers were continually honing their design and construction skills.

Through their work at the Wright Cycle Company, they were able to intrinsically understand three key elements needed for building an airplane.

* First, experimentation was the key, just like it was in designing a bicycle. A design on paper may not work well when built or tested.
* Second, recognition that even with machine-powered flight, the person in the plane – Wilbur or Orville – would need to learn how to use their weight to help the plane change direction, the way one does on a bicycle when leaning into a turn.
* Third, learning to fly a plane, just like riding a bike, was going to take practice. Lots and lots of practice, with the ever-present risk of falling, and accepting that injury was a price one must pay in the learning process. Every flight would involve risk.

And there was something else they agreed on. All their research, work and testing would need to happen during their free time, when they were not running the bicycle shop. They had recently launched two Wright brothers' designed bicycles: the St. Clair and the Van Cleve. Both were popular.

For this to work, the men agreed they would need to be extremely efficient with their time and run a financially sound business, to ensure a portion of the shop's profits could be used to buy materials to build kites, gliders, and ultimately airplanes.

May 1899 – the start

Based on the research they had done on birds and birds in flight, Wilbur and Orville felt they could contribute to efforts being taken in multiple nations to construct a viable heavier-than-air flying machine.

Wilbur wrote to the Smithsonian Institute in May 1899 explaining how they had tried building larger examples of Cayley's and Penaud's toy helicopters, and that the brothers were serious in their study of flight. In the letter, they requested books and materials be sent to them so they could study. The Smithsonian sent the books, and Wilbur and Orville now became more fully aware of the works and writings of Langley and Chanute.

The brothers started building box kites, which had an upper wing and a lower wing beneath it. Struts separated the two wings, creating the open-sided box shape. Their design was similar in structure to Lilienthal's glider. As every child with a simple kite learns, the problem is not launching the kite but keeping it in the air. Wilbur and Orville experimented with various designs of box kites, combining this with what they had observed of birds in flight.

To describe the Wright brothers as *merely* a pair of bicycle shop owners is to do them a grave disservice. They acted like the most serious scientists, approaching each test flight as a controlled, documented, and photographed experiment. They quickly excelled at assessing and analyzing their results and learning from the successes, failures and setbacks.

They were not afraid to challenge conventional wisdom.

By examining Lilienthal's glider flights and how the pilot's legs dangled below the frame, they realized that not only was Lilienthal's position extremely precarious, but that controlling the glider's movement solely through the pilot shifting their body weight was inefficient.

Wilbur and Orville went back to watching birds. What did birds do when they wanted to turn? They surmised that birds flick the very ends of their wings. It's a subtle motion that has a profound effect.

The question became how do you mimic this in a kite or glider?

The answer was not building a rigid structure but one that could flex. This was where their real genius showed through. They had the ability to see a problem, analyze it rigorously, identify what the solution could be, and then apply creative ingenuity to build it with just the tools in their bicycle shop.

Using string and a more flexible wing design for the kite, they developed a way for the tip of one wing to move upwards while the other side of the same wing tipped downwards.

They called this Wing Warping.

Tests in the fields of Dayton, Ohio proved that their design was starting to give them the control they desired.

Several local children watched Wilbur fly the guided kite and marveled at how it floated in the air turning and lifting. That was until Wilbur lost control, the kite strings went slack and the children dove for the ground as the out-of-control kite smashed into the earth.

It would take a lot more tries to perfect Wing Warping. The challenge of building a real flying machine was still astronomically enormous.

Many questions remained to be answered:

* How do you translate a design for a kite into a glider large enough to support a human being?
* How do you enable the pilot of that glider to control the wing warping from their central position on the glider?

★ And how do you build that control mechanism to be light weight, strong and unbreakable, given that any margin of error could mean certain death or serious injury to the pilot?

The Wright brothers made two other breakthroughs in their kite experiments. The first was to attach a controllable front rudder, which they called an "elevator," to the front of the kite. The second was constructing the wings to be curved and angled so that if you looked at a cross-section, the highest point was not at the center point but much closer to the front of the wing.

Kitty Hawk

In the summer of 1900, Wilbur and Orville wanted to experiment with a full-sized glider. By applying Lilienthal's tables, they concluded their two-winged biplane glider would need both wings to have a minimum combined surface area of 150 square feet (14 square meters) if flown in a consistently windy area, delivering 16 mile per hour (25 km per hour) steady winds.

Dayton was just not windy enough, so Wilbur wrote to Chanute – the leading aviation inventor of the time – explaining their work and the need for a glider test location. Chanute's reputation meant he was constantly approached by governments, universities, scientists, and inventors, as well as amateurs, cranks and charlatans, all seeking information or wanting to tout their latest flying invention. Chanute could tell from Wilbur's letter that he was not only serious but already solving real flying dilemmas.

In Chanute's reply, he recognized that although California and Florida might have the wind, the Wright brothers should seek a place that had sand and sand dunes for softer, safer landings. He recommended Georgia or South Carolina. Wilbur and Orville then sought US Weather Bureau reports about windy places and settled on the remote location of Kitty Hawk, North Carolina, which was one of the windiest places in America.

Kitty Hawk was also scorching in the summer, bitterly cold in the winter, and plagued with hungry mosquitoes. In an era that was pre-sunscreen and pre-insect repellent, one can only imagine the discomfort that befell the brothers. Add to that the difficulty of being in an extremely windy place built on sand. The wind had the ability to whip sand around, turning it into biting particles.

But none of this bothered the brothers. And this was one of the most remarkable traits we found in the very small teams we studied. Extreme physical discomfort was but a minor annoyance in the pursuit of epic goals.

This was as true for Wilbur and Orville Wright as it was for many of the other very small teams you will meet in this book, all of whom were more dedicated to solving the problems at hand than they were to notions of personal safety and comfort.

Though Kitty Hawk was sparsely populated, the town officials and locals were hospitable and interested. The most notable among them was William Tate, a local farmer who also served as Kitty Hawk's postmaster. Tate gave Wilbur advice about suitable locations, weather conditions, directions, campsites and who among the locals may be willing to provide him with hot food.

Splitting up the tasks

Working together but splitting up the tasks, the Wright brothers built their first glider in their Dayton shop. Its wings measured 18 feet (5.4m) long and 5 feet (1.5m) wide, with the upper wing 5 feet (1.5m) above the lower. The wings were connected by wood struts with flexible joints, braced with metal cables.

Steel controlling cables and pulleys were ingeniously embedded into the wing structure to enable the operator (the Wright brothers did not use the word pilot at that time) to warp the wings, and thus control flight. The glider had an elevator (rudder) projecting from the front, to help control altitude.

Rather than using wheels for the undercarriage, it used skids that they reasoned would work better on the Kitty Hawk sand. The only pieces missing were some 18-foot (5.4 m) spruce spars for the wings. That wood was unavailable in Dayton. Once built, they disassembled the glider in Dayton, and crated the parts. Wilbur traveled with it by train to North Carolina hoping to buy the spruce spars there, but was only able to obtain 17-foot (5.2m) spars made from pine not spruce.

Befriended by William Tate and his wife, Wilbur set up an encampment in Kitty Hawk and re-assembled the glider using the shorter pine spars. In the process he had to re-sew the wing material to fit the smaller sized wings – a perfect example of the inherent craftsmanship that Wilbur possessed. He could build, adjust, and modify designs quickly in the field using the tools at hand.

Orville joined Wilbur a few weeks later. After all, they still had a bicycle shop to run. Orville's journey to Kitty Hawk was considerably easier. For one, his baggage didn't include the parts of a glider.

Orville was immediately struck by how desolate Kitty Hawk was. He described that trying to find Wilbur across the windswept sands must have been as difficult as what Arctic explorers experienced when seeking a lost companion.

Their first glider

Their plan was to fly the glider as a kite from the ground. This would give them the needed experience to learn how to control it in the air. But in the strong winds, approaching 30 miles per hour (48 km per hour), they found control difficult. The glider wanted to soar higher. The challenge was safely preventing that without the glider crashing the way Wilbur's box kite had in the Dayton, Ohio test.

They were not always successful.

In one test, while controlling the glider/kite from the ground, it suddenly smashed to the ground flinging Orville, who was holding one of controlling ropes, quite a distance. Fortunately, he was unhurt, and the glider was soon repaired and ready to fly again.

In a very small town of working people, Wilbur and Orville, who dressed in suits and ties regardless of the weather, were quite a sight. The local people were as amazed by their diligence, perseverance, and incredible work intensity as they were at what they were trying to achieve.

Once the unmanned trials were completed, the manned trials were next.

Local townspeople came out to watch Wilbur lying prone in the center of the glider, marveling at how he managed to keep the glider under control while in the air and then safely bringing it down onto the sand.

Wilbur Wright in prone position on glider just after landing, its skid marks visible behind it and, in the foreground, skid marks from a previous landing; Kitty Hawk, North Carolina. *(Library of Congress)*

Upon analyzing the results of their many trials, Wilbur and Orville found that the glider flew well in very strong winds like Kitty Hawk's 25 to 30 miles per hour (40-48 km per hour) ones, but at lower wind speeds, it did not have the expected lift predicted by Lilienthal's tables for the 190 pound (86 km) combined weight of man and machine.

Was their design at fault, or were Lilienthal's tables wrong?

The pair undertook detailed analysis and concluded the famous man's tables were indeed wrong. The bicycle builders from Dayton with only one high school diploma between them had just disproved one of the fundamental tenets of the aviation world.

During their last days in Kitty Hawk in October 1900 and with the help of William Tate, they took their glider up to Big Hill, the tallest of a set of large sand dunes. Launching from this height gave Wilbur a considerable advantage.

He successfully made flights reaching altitudes of 300 to 400 feet (91 to 122 m), and speeds of 30 miles per hour (48 km per hour).

To Wilbur it felt safe, controllable, and most of all, it felt like flying.

Chapter 7
Flying Machines

"Not In A Thousand Years"

Knowing they needed to create their own tables to replace
Lilienthal's, they spent the winter of 1900 and spring of 1901
in Dayton planning and building a bigger glider. With the
knowledge they had gained thus far, they had more data to share
with Chanute, who visited them in June 1901.

Leaving Wright Cycle Company in the capable hands of Charlie
Taylor (who will reappear in an important way later in this story),
Wilbur and Orville returned to Kitty Hawk in the summer. They
set up camp, dug a fresh water well, built a larger structure to house
the glider, and then the assembled the glider itself.

They never stopped working, even while enduring extremes
of conditions: blistering summer heat; vicious mosquitoes biting
through their clothing during the day, and even worse at night;
merciless windblown sand; and bouts of driving rain. Some days
there was not enough wind and other days far too much, to safely
control the glider.

They persevered, only to find that their new bigger glider, which
had a 22-foot (6.7 m) wingspan and other modifications, performed
worse than the first. With it they could only achieve very short
flights, some ending in injury-causing nosedives into the sand. The
brothers changed the curvature of the wings to be closer in design to
their first glider. They experimented with flying the glider as a kite,
and sometimes with bags of sand in place of an operator to assess

how the craft responded to different weights. And, on occasion, one of the brothers piloted the glider.

They made hundreds of glider flights and took detailed measurements of altitude, wind speed, and distance achieved – which could be as far as 400 feet (122 m). They carefully and systemically adjusted the pulleys, brace wires and struts, and altered the size and position of the elevator, taking note of which changes made improvements and which did not. But overall, they were disappointed and discouraged with their progress.

It felt like they had taken a big step backward. Wilbur at one point even exclaimed that he didn't think a flying machine could be built in the next one thousand years.

Wind tunnel

One of the advantages of being part of a very small team rather than working individually is having a teammate to buoy your spirits. Setbacks and discouragement are to be expected when pursuing great goals, but it takes great teamwork to bounce back stronger.

Achieving success requires that when one team member is feeling down, the others boost their spirits. And this is true, whether the small team is doing physical adventures in the Polar regions or climbing Everest, or doing more cerebral endeavors like writing musical theatre, designing buildings, or changing the political landscape. It's equally true when designing the first airplane.

For the Wright brothers, the answer to overcome their setback was simple.

Work harder.

They knew the biggest variable in assessing what adjustments to the wings, struts and other parts of the glider would be needed to improve lift and maneuverability was the wind. It rarely blew at a constant rate or from a single direction in the open-air flying tests at Kitty Hawk. So after Wilbur and Orville returned to Dayton, they designed and built their own wind tunnel to test scaled-down models of their glider and its wing formations.

Their first wind tunnel was too small.

Undaunted, they built a bigger one and designed a device to measure lift and drift of the model glider. For months they undertook painstaking and detailed tests.

Throughout, their partnership was rock solid. They still lived in the same house with their sister Katherine, who ran the household. But it wasn't all plain sailing.

Wilbur and Orville didn't always agree and could have loud and boisterous arguments about their work, the experiments, and what the outcome of their analysis meant for glider and wing design. At times as arguments went on, they could even end up switching sides, each arguing against the point they were previously defending. This gave them an added advantage – two people, with complementary and nearly interchangeable skills, focused on the same problem at the same time seeking the same, exact goal.

Knowing that all previous tables by experts about lift, drift, pressures on wings, locations where the center of pressure occurred, and other key metrics were seriously flawed, they set about creating new, precise tables derived from their experiments. At this moment, in their workshop above their bicycle shop, working part-time on the flying problem, and despite their lack of university degrees and not seeking or needing external funding (although Chanute had made multiple offers of money), Wilbur and Orville were emerging as the preeminent aeronautical engineers of their generation.

The remarkableness of this very small team cannot be understated.

Newer and bigger glider

With their new tables, they returned to Kitty Hawk in the summer of 1902 to assemble a glider with an even longer wingspan of 32 feet (9.7 m). It had a larger, adjustable front elevator and fixed vertical vanes on the rear. They kept the same style of pulleys and wires to create the crucial wing warping as they had on their previous glider.

Early experiments with this new glider showed some advantages, but the operator, whether it was Wilbur or Orville, still struggled to control it. After more experimentation and adjustments, they had a breakthrough and turned the rear vertical vanes into a steerable rudder.

But this simple change had a downside. It made the operator's job even more complicated and prone to error, because there were more movable parts to control and adjust on take-off, mid-flight, and landing.

To solve this, the brothers ingeniously figured a way of linking the wing warping and the rear vertical vanes to move simultaneously and automatically, which enabled a stable glide.

Their ability as a team to see a problem, identify a solution, implement that solution, and then risk their lives testing it over and over again might be unprecedented. In that summer they performed over 700 test flights of the glider. The Wright brothers now had the wing and navigational design.

But that was for glider flight only. The next challenge was to add an engine.

Both knew an internal combustion engine would be heavy, so while still at Kitty Hawk, they reinforced the glider with additional struts and trusses, and test-flew it with added weight. With their recalculated lift tables, they could determine that the end-to-end wingspan would need to increase to 40.5 feet (12.3 m).

The brothers returned to Dayton to now solve an even bigger challenge: the engine.

But another challenge also loomed.

Other better-funded, serious and experienced inventors were trying build flying machines. Who would be first?

Adding a motor and propellers

In 1902, automobile engines were being built by many companies, but none were designed to be lightweight. Because of this, the Wright brothers would have to build their own.

Even though neither man had designed an engine before, they studied the problem as a team, focusing on how to build a lightweight, yet powerful engine that could drive propellers, and then set about designing one.

It was at this point that Charlie Taylor, the man who stayed behind in Dayton to look after the Wright Cycle Company, became a vital, unsung hero in the creation of first successful flying machine – almost an adjunct member of the Wright brothers' team. Charlie, a talented mechanic, had previously worked in manufacturing farm machinery and had rebuilt a car engine once.

The Wright brothers commissioned an aluminum manufacturing company to build an engine block to their specifications, using the profits from the bicycle shop. But an engine block is not an engine. Using the metal lathe and other tools in the shop, Charlie built the pistons, camshaft, valves and lifters, a gas tank, and every other piece to turn the block of aluminum into a working engine.

During the first tests, the engine block cracked. Had this been mounted in the glider, there would have been a crash landing and a dead pilot. Undaunted, the brothers ordered a second block and Charlie built a second engine.

His complete engine came in at 48 pounds (21.7 kg) lighter than originally expected and could generate 12 horsepower compared to the 8 horsepower they originally thought it would. This was a great achievement but there was still another big hurdle: the propellers.

They had to design these too.

Thinking propellers would be designed like those on an ocean-going ship, the brothers quickly realized that was a wrong assumption. Using their wind tunnel to analyze the airflow over moving vertical surfaces, they designed the first airplane propellers ever built at that time.

But where should the propellers and the engine be fitted in the flying frame? And, how can they be controlled?

Through experimentation on the ground, they worked out one propeller would need to spin clockwise while the other spun counterclockwise, to balance the plane's movement.

The brothers would have one chance to get it right in the air. They already knew how an out-of-control glider accident in the winds and sands of Kitty Hawk could cause a bone jarring injury. They had experienced that many times.

Whether it would be Wilbur or Orville who piloted that first test flight, they would be risking their life in an entirely different way.

The previous plane was just a glider. A gasoline-engine-powered aircraft with rapidly spinning propellers turned by bicycle chains, all of which was right next to the pilot lying prone and unprotected between them, involved a much greater risk. Error at a faster speed meant death or a life-changing permanent disfigurement.

Kitty Hawk, 1903

As in the previous years the brothers arrived in Kitty Hawk with the glider in numbered crates and had to start by establishing a camp with an even larger outbuilding (what we might now call a hangar). Only then could they start assembling the plane. All this took place from September to early December 1903.

The new glider, named the *Flyer*,[5] had several features including a new and improved rib design for added flexibility and strength. According to Orville's later writings, this rib design was used in almost every airplane built for the next ten to twenty years. In another innovation, the wings were double clothed, with fabric covering both the top and underneath surfaces of each wing.

Hampered by mechanical challenges and blustery cold and rainy weather, it was already December 14th when they were ready to test the *Flyer*. And which brother would pilot the craft had yet to be determined.

They flipped a coin. Wilbur won; he had the first go at flying it.

A few local men watched Orville run alongside the plane as it accelerated along a track and lifted into the air, only to see it crash to earth. Unfamiliar with the yet-undiscovered challenges of engine-powered flight, Wilbur had overcompensated on the lift off. Wilbur was physically fine, but the plane needed some repairs.

It took three days for them to get the plane ready for the next attempt. Now it was Orville's turn.

They set up a camera and instructed John T. Daniels, one of the

---

[5] Also known as the *Wright Flyer*, 1903 *Flyer*, *Flyer 1*, and the *Kitty Hawk Flyer*.

five local men who came to watch the flight, to press the shutter button when the plane lifted into the air.

Exactly as was done a few days before, the plane was brought into position on the track. Only this time, Orville and Wilbur stood together near the plane and shook hands for a longer time than would be customary. One of the local men described it as looking like two men who might never see one another again.

Orville lay in a prone position on the plane. With a whirr and a roar, the engine and propellers were started, and the *Flyer* began to move. Wilbur ran alongside, holding onto a wingtip as the aircraft accelerated down the track.

The plane lifted into the air. It travelled 120 feet (36.5 m), flying 10 feet (3 m) above the ground, and then Orville successfully landed it. The entire flight lasted 12 seconds.

The first flight. Orville flying the plane; Wilbur running alongside.
(Photograph taken by John T. Daniels.) (Library of Congress)

The photograph was perfectly framed. It shows the *Flyer* in the air, Orville lying prone within the plane flying it, and Wilbur running alongside.

It has become one of the most famous photographs ever taken. It shows not just the first flight. It shows the enduring power of a very small team.

John T. Daniels, who had pressed the shutter button at just the right moment of that short 12 second flight, had never taken a photograph before in his life.

A team

In little more than three-and-a-half years, from a near standing start at looking seriously at the challenge of flight, inspired from the time when Lilienthal died in his crash, the Wright brothers had done what many, including the greatest minds in the world – da Vinci, Galileo, Edison, and others – had dreamed of for centuries.

They had conquered flight.

They did this while working on the problem part-time, totally funded with their own money. They did this from their bicycle shop and from wind-swept plains of Kitty Hawk, North Carolina.

And they did this as a team.

★

Conquering flight took a special kind of bravery, which might be likened to the bravery of polar explorers. The next story is about Peary and Henson. They risked their lives over decades to become the first to trek over sea ice and stand at the North Pole.

Their story, and especially the ending, will likely startle you.

Robert Peary on the main deck of steamship SS *Roosevelt*, 1909.
Matthew Henson, 1910. *(Libray of Congress)*

Chapter 8
Exploration

Peary and Henson

"It'll work if God, wind, leads, ice, snow, and the hells
of this damned frozen land are willing."

—*Matthew Henson,*

Exploration has always been a team endeavor.

The innate desire of humankind to extend its knowledge of the Earth and uncover its deepest secrets cannot be fulfilled by the actions of one person. The journey to explore the unknown will always be a daunting proposition. It requires far more than curiosity and desire. It needs an iron will, boundless endurance, and a limitless urge to quell that human curiosity. It requires a team effort.

Our focus now turns to one of the longest serving polar exploration teams that ever set out to quench that desire: the American explorers Robert E. Peary and Matthew Henson.

The Peary-Henson story began in 1887 and drew to a close immediately after their 1909 quest to be the first men to reach the North Pole. Along the way, they proved that great endeavors can be done by two people with a common mind and will, and who are committed to strive for an unlikely, almost unattainable goal.

Every other acclaimed polar leader of this era, including the Norwegian heroes Fridtjof Nansen and Roald Amundsen, the leaders of the British Antarctic expeditions Robert Falcon Scott and Ernest Shackleton, and the intrepid Australian explorer Douglas Mawson – all only took white men on their expeditions.[6]

Robert E. Peary was different.

In an era where no women were taken on expeditions, Peary took his wife, Josephine. She joined him on several of his Greenland expeditions. And no expedition leader of that era had ever taken a Black man on his expedition until Peary took Matthew Henson. Not only was Henson a Black man in the white world of polar exploration, Peary chose him to be his right-hand man for every Arctic-bound expedition.

But to portray Peary as an enlightened man of his times who could see beyond race or gender in his pursuit of the perfect polar team is to give him credit he may not deserve. This is a question to contemplate when reading the next chapters.

Matthew Henson became Peary's most valued team member, and in the end stood side-by-side with him at their last northernmost camp. In pursuit of their common goal they were a team – or so Henson believed. And that is the crux of this story.

The start of the great adventure

By 1900, most of the world had been visited, explored, and mapped, but a few of the most forbidding regions remained untouched, untrodden by the boots of men, and unnamed. At that time more was known about the moon than the Earth's Polar Regions, but these gaps in knowledge were fast disappearing, driven by curiosity and the twin greeds of fame and fortune.

These were heady times for explorers. The Antarctic continent at the bottom of the planet had been seen and even walked on in places along its perimeter, but no one had yet gone far enough inland to

---

[6] How the early Antarctic explorers chose their teams is described in our first book, *When Your Life Depends on It: Extreme Decision Making Lessons from the Antarctic*.

describe the lay of the land, or to seek out the route to the South Pole.

The same could be said for the Arctic in the North. Most of the northern continental shores of Europe, Asia, and the Americas had been charted. What lay beyond – to the north – was barely discernible fields of floating sea ice as far as the eye could see.

No one knew precisely what would be found there. Could there be archipelagos of mineral-rich islands, or perhaps another continent? Repeated forays into the icy wilderness extending existing records of "farthest North" had netted little of commercial or colonial value. They were achieved at the expense of great effort, much misery, and sometimes even death for the explorers. On the plus side, they had brought acclaim for the leaders of those expeditions, and delivered a boost to the egos to the rulers of the nations who had sent them.

American explorer Robert Peary had long desired to be known as the discoverer of the North Pole, the one man who would succeed where many other famed explorers had failed. From his early childhood, Peary sought that fame with all his might. Over the course of eight expeditions he expanded on own his own "farthest Norths" in pursuit of his ultimate goal: the North Pole.

Such expeditions needed tons of supplies, many men to move them, a ship to carry them north, and money to pay for it all. Peary was a consummate organizer, a finder of money and supporters, a man driven by a thirst for adventure, and hungry for fame. But he also needed personal support, people in his life upon whom he could depend, while he chased his own dreams. He found it in his wife Josephine, and his steadfast expedition companion, Matthew Henson.

In those days, Black men were never welcomed into the ranks of these brave explorers. In the late 1800s and early 1900s, exploring was an expensive enterprise sponsored by wealthy businessmen who met in luxurious, private gentlemen's clubs, a place where people of color were allowed only as servants. In that world, Henson could never dream to be more than Peary's assistant. He lived in a United States where Blacks were paid less, and treated worse than their white counterparts for doing the same job. In that respect, Peary's expeditions were no different.

Was Henson – who shared with him the inevitable extremes of cold and hunger in that forbidding climate – never much more than a supremely skilled servant to Peary? Wouldn't Peary and Henson have to be a team if they shared all the toils and triumphs of those expeditions? What role does the leader of a very small team play when risk and danger are all around?

These are the questions that make the Peary-Henson relationship so fascinating.

Peary's plans before meeting Henson

In 1881 Peary was working for the Survey Department of the United States Navy when he saw an opportunity for career advancement. The Navy was seeking a chief surveyor to determine if Nicaragua would be preferable to Panama for a canal connecting the Atlantic and Pacific Oceans.[7]

Peary applied and four years later in 1885 made his first foray into the stifling rainforests of Nicaragua. When no viable route could be found, Peary knew he would have to go back again, but during this time, a powerful yearning to pursue Arctic adventures had arisen within him.

Although Peary had no practical experience in polar exploration, he nonetheless developed a plan to explore Greenland in the following year, during periods of extended paid leave from the Navy. His ideas and plans were derived from the study of literature covering all prior expeditions, from which he devised a simple rule – the more men and equipment he brought along, the less likely an expedition would achieve its desired outcomes.

Peary believed success would come from having one leader – Peary of course – with a small team of men, "selected solely for their courage, determination, physical strength, and devotion to the leader."

---

[7] The history of the 400-year quest to build a canal across Panama, and the United States' interest in selecting Nicaragua as the crossing point rather than Panama is explained in our book, *Audacious Goals, Remarkable Results*.

This meant a greater reliance on the local workforce – the native Inuit[8] hunters who, with their wives and especially their dog teams, would provide sledges, meat and fur clothing, and the power to haul it all into the unexplored icy terrain of northern Greenland. From Peary's point of view, this made sense logistically and economically, since these services could be purchased for as little as a trader's assortment of metal-based objects: knives, needles, guns and ammunition.

In 1886, using a gift of $500 from his mother, Peary bought supplies and booked passage north on a ship to Godhavn, Greenland. He had originally intended it to be a solo trek until Christian Maigaard, a young Danish official based there, convinced him that an inexperienced man going alone would be suicidal.

Maigaard teamed up with Peary.

The pair trekked 100 miles (160 km) inland before being forced to turn back due to lack of equipment and food. Even so, this was the farthest penetration onto the Greenland ice sheet at the time. While Peary was satisfied with the results of this risky journey, he knew the next attempt would have to be more serious.

Beyond proving that use of a small team was better and placing his name on notice as a polar explorer, Peary's foray into Northwest Greenland did little else.

The hat shop

Before making another trip to the Arctic, Peary had employment obligations and needed to return to Nicaragua.

The following year in 1887, when Peary was gearing up for this second Nicaraguan survey, he walked into a haberdashery in Washington, D.C. looking to buy a pith helmet. There, the owner introduced him to one of the shop assistants, 18-year-old Matthew Henson.

Peary was seeking more than jungle headgear. He was also

---

[8] The term "Inuit," meaning "the people" in the Inuktut language used by the indigenous people of the Arctic, has replaced the pejorative term Eskimo once commonly used by outsiders before 1980.

looking for a valet – or as Peary put it, a 'body-servant' – to look after his clothing and to tend to his domestic needs. At that time, it was a job that frequently went to Black men.

Matthew Henson had been born into a recently freed Black family in Maryland in 1868, only three years after the end of the Civil War and the end of slavery in the United States. During Henson's adolescence, there were no good career paths for young Black men. Aspirations were regularly crushed by the weight of deep-rooted racism. Henson fled an abusive stepmother at the age of 13 and walked to Baltimore in the hope of finding a berth on a ship there. He found sympathy in a man called Childs, captain of the outbound China clipper *Katie Hines*, who took him on as a cabin boy.

Young Henson used this opportunity to impress the captain with his adaptability and intelligence. Captain Childs in turn taught him the trades of the sea, to read and write, and perhaps even the rudiments of navigation. After three voyages to China, Henson took up residence in Washington D.C., working at various trades including the haberdashery.

Henson listened to the opportunity that Peary described. It was too good to ignore. Peary hired the young man first as a valet, then promoting him to a valued member of his Nicaraguan survey team, where Henson quickly proved his versatility.

Well before their time in Nicaragua ended, Peary appointed Henson as his chief 'chain man,' – a key position in the survey party. But despite Peary and Henson's efforts, the survey team could not find a viable canal route.

During that year in Nicaragua, Henson had proven to be indispensable to his boss. They had established a workable balance between leader and follower. To understand this partnership, one must read between the lines of Peary's ghost-written published works, derived from his diaries and unpublished letters, and Henson's ghost-written book, as well as a few surviving personal documents. These two men chose each other's company time and again over the course of decades to navigate the most inhospitable places in the world.

Neither might have confessed to the fact that they needed one another, but they surely did.

Peary needed a versatile and dependable dog-driver, hunter, and Inuit language translator – a language Henson would learn during his years on the expeditions. Henson needed a job, a way to make a living doing something that recognized his talents and satisfied his own taste for adventure.

In light of their backgrounds, their working relationship made sense to them, and at that time it made their work possible. This is the essence – or one of them – of teamwork. Although the members may have different skill sets and different roles, and even differing statuses within the team, they still work together in pursuit of a common goal.

The money and the acclaim for polar success would go to Peary. Henson would receive a wage while on each expedition, and when he returned to the United States, Henson would go back to scrounging for work wherever he could find it.

During Peary and Henson's years together in the Arctic pushing past the boundaries of the known world, they sought and suffered for the same goals with equal fervor. Their skills during this novel enterprise improved in parallel. Before traveling to the Arctic, neither had driven a dog sled, built an igloo, suffered frostbite, known hunger to the point of starvation, or spoken with or worked alongside the Inuit people who called this place home.

But during their time in hostile and cold environments, they also discovered new lands and pioneered new techniques of exploration. The impetus behind this activity was Peary's personal quest for fame and security. Henson would never have come here otherwise. He could have pursued other employment, but chose to go with Peary every time the opportunity arose.

Peary's goals became his goals.

Chapter 9
Exploration

Striving Northward

The intentions of Peary's early Arctic expeditions were to garner him a little fame for having explored northern Greenland. The western shores of Greenland were already well known – Baffin Bay had been a regular haunt for whalers for centuries. The east coast of Greenland was not as well suited to this marketable commodity, and consequently less well known. Greenland's northern part was entirely unknown and unexplored.

It literally was *terra incognita*.

Important questions awaited answers.

Was Greenland an island or the southernmost extension of an undiscovered continent extending to the North Pole and beyond? Were there exploitable natural resources there?

Peary wanted to be the one to find out.

After the Norwegian explorer Fridtjof Nansen's 1888 success in crossing the southern Greenland ice sheet, Peary, unable to bear the thought of another explorer trespassing on "his" turf, immediately began planning his first serious expedition.

His goal was to cross Greenland to the northeast limit of the Atlantic side, while mapping the unknown interior along the way, to find an answer to his first question. Such a discovery would significantly alter how a map of the Earth was drawn and how the North Pole could be conquered.

To find out, Peary's first fully-funded, major Arctic expedition began in 1891. A steam sealer ship *Kite* dropped off the wintering party of seven on the shore of McCormick Bay, in Northwest Greenland. Among the team were Peary's wife Josephine, Henson, and Dr. Frederick Cook – a medical doctor on this expedition who would later become Peary's arch rival.

Josephine Peary's inclusion meant that she would be the first woman to participate in a polar expedition. Thanks to the pioneering work of Susan B. Anthony and Elizabeth Cady Stanton, this was a time when women were feeling more empowered. By 1891, women were achieving more rights and freedoms every year.

This was also Henson's first expedition to the Arctic. Peary had already come to depend on Henson's ingenuity and loyalty. Henson's value was left in no doubt when Peary entrusted to him the protection of Josephine at their base camp.

Peary hired local Inuit families to work for him. The husbands to hunt and teach the white men how to drive their dog teams over the ice; the wives to sew furs and keep them company. His successive expeditions would come to depend increasingly on their skills and services.

Despite the many years Peary spent in their company, the mutual respect between him and the Inuits never blossomed into friendship. Henson, in contrast, learned their native language and their ways, and by the end, was accepted into their communities, more as an equal than a superior.

Peary's plan was to leave Josephine and Henson at the base camp, while he and twenty-year-old Eivind Astrup headed across the ice. Theirs was a gruelling 1,250 miles (2,000 km) journey. They reached the north-eastern shore, and in doing so, became the first to cross the northern ice sheet from one side of Greenland to the other, and to name its geographical features. Standing at the edge of the vertical Navy Cliff, at the most northern part of Greenland, Peary named the deep gorge below them Independence Fjord in honor of the day – July 4, 1892. But their excitement was short lived as supplies were running low.

With only six days of food left, Peary and Astrup barely made it

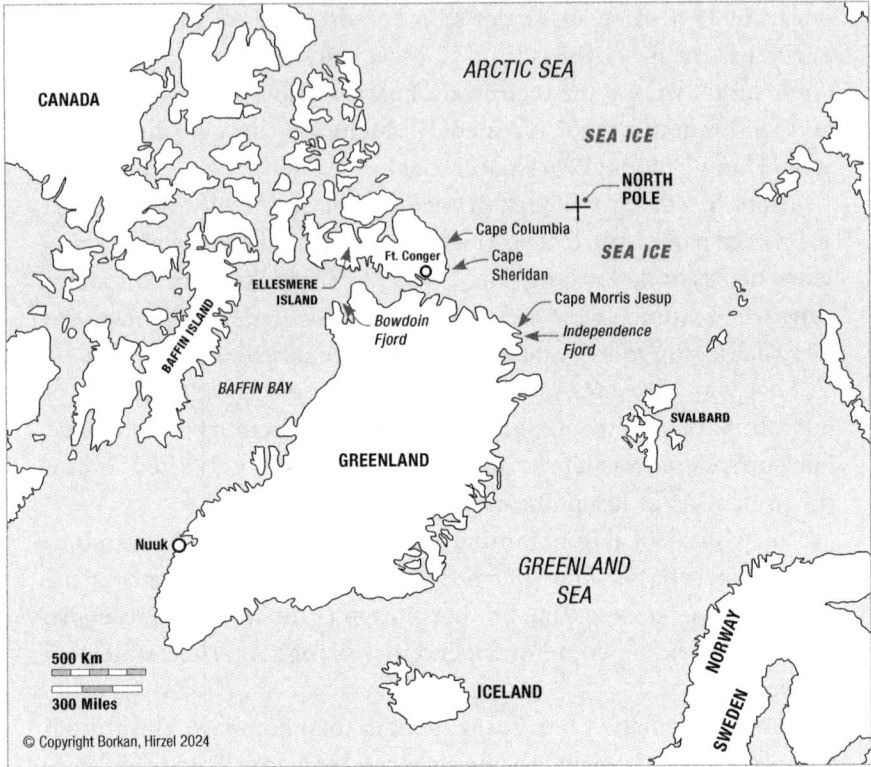

Greenland, the Arctic and the North Pole. (*Locations and distances are approximate.*)

back alive to their base camp on the west side of Greenland. Only one dog remained; the rest they had eaten along the way. But this very small team had proven something important: Greenland was indeed an island, and not the southern end of a great North Polar continent.

Compared to Nansen's southern Greenland traverse which took 40 days four years earlier, Peary and Astrup on this northern route trekked four times Nansen's distance in 97 days.

After these tantalizing results, Peary wanted more.

Lecturing and getting ready for the next expedition.

Peary knew that to travel farther in the Arctic he would need more time and money, and a better organization. Building on the fame of this expedition, he embarked on a lecture tour. To enliven his talks, he brought Henson along, not for the man to describe his experiences to the audience but to model the Inuit clothing they had used. Henson's other duty on these tours was to manage and display a dog sled team, which became an attraction of the event.

While Peary and Henson were busy on the lecture circuit in the United States, Astrup gave a lecture about the expedition in his home country of Norway. A 20-year-old Roald Amundsen, already filled with his own polar ambitions, was enthralled by Astrup's talk. Years later, Amundsen gained acclaim for being the first to discover the Northwest Passage, the first to trek to the South Pole, and the first to fly over the North Pole[9].

The Peary-Astrup crossing of Greenland had given Peary a taste of the real fame that would accrue to the man who would be the first to stand on the North Pole – that invisible axis upon which the Earth rotated. Peary wanted to go back to Baffin Bay, retrace his steps, and scout a route ever farther North – a route that could lead to the Pole. Henson agreed. Not just for the work, but because he was beginning to share his employer's thirst for discovery.

Peary began organizing this next expedition. Henson's role was, as always, that of a helper and not a decider or equal. Henson's real work did not really begin until the ship was at the dock being loaded for the expedition. Here he was in his element. He brought to this task his inherent love of order and efficiencies – skills always in demand when loading a ship for departure.

Eivind Astrup and Josephine Peary were again on board when the *Kite* departed for the North in 1893. Josephine was pregnant with Peary's child. In September, their daughter, Marie Ahnighito Peary, was born in the hut called *Anniversary Lodge* on the shore of Bowdoin Fjord, in northern Greenland. The Inuit women came to visit Josephine and her "Snow Baby," the first white child they had ever seen.

---

[9] Roald Amundsen's achievements are described in our book, *Audacious Goals, Remarkable Results: How an Explorer, an Engineer and a Statesman Shaped our Modern World.*

Peary's goal of finding a route to the North Pole was not achieved, despite attempts in both 1894 and 1895.

Fort Conger

Peary's next venture in quest of the North Pole began a few years later, when the ship *Windward* dropped off Peary, Henson, the expedition surgeon Dr. T. S. Dedrick, and their Inuit companions on the East coast of Ellesmere Island. Trekking overland north to Fort Conger, they arrived there on January 6, 1898 and set about dismantling the remains of the old hut left by a previous Arctic expedition team. They used the lumber to build three smaller huts, low to the ground and interconnected by trenches through the deep snow that were roofed over with hides and canvas. They would spend the next four winters at Fort Conger, from 1898 to 1902.

Here was where Henson came into his own as a polar explorer.

Peary later said of him, "Henson can handle a sledge better, and is probably a better dog driver than any man living, except some of the best Eskimo hunter themselves." Yet, these skills did not merit any more pay than $50 per month, less than Peary paid the white men doing the same work, but better than Henson could find as a Black man back home in Baltimore.

Peary, engrossed in the heavy work of setting up the base, did not notice the growing numbness in his feet. By the time he did, nine of his toes had been damaged by frostbite. Dr. Dedrick, with Henson as his surgical assistant, amputated them by lamplight in the dim, dark confines of one hut. Although Peary recovered enough to continue sledging the following summer, he was compelled to travel as a passenger rather than a driver and walked with what Henson described as a peculiar sliding gait for the rest of his life.

Undeterred, Peary, Henson, Dr. Dedrick and their Inuit sled drivers crossed over to Greenland in the spring, seeking as always the farthest northern shore from which to begin a journey to the North Pole. Hunting down musk-oxen along the way to augment their food supply, they reached Greenland's northernmost point at latitude 83° 40'. Peary named it Cape Morris Jesup after one of his wealthy sponsors, and recognized it a potential launch point to start a trek over the sea ice to the North Pole. But once on the ice, they only gained a few more miles, to secure a new "Farthest North" of 83° 50' before heading back to Fort Conger.

It was during this time in 1900 that Josephine planned a surprise visit to Fort Conger. She brought their daughter along. Her ship, which was damaged in a collision with an iceberg, had to take refuge for the winter on the Greenland shore, 300 miles (482 km) south. There Josephine was astonished to meet an Inuit woman named Aleqasina who innocently boasted that the young male child with her was Peary's. Years later Henson acknowledged that he too had an Inuit "country wife" who bore his child.

Joined by the hired Inuit and their dog teams, Peary and Henson headed northwest along the unexplored northern shore of Ellesmere Island beyond Cape Sheridan, and eventually located the northernmost point of Ellesmere Island, naming it Cape Columbia. Although not quite as far north as Cape Morris Jesup, it was nonetheless better situated for sledging supplies overland from the nearest possible ship anchorage. With only sea ice between Cape Columbia and the North Pole, this place became the starting point of leaving solid land to trek across floating slabs of sea ice for all of Peary and Henson's expeditions that would follow.

A first big push out over the sea ice

The going would not be easy, but wherever Peary went Matthew Henson went with him, oftentimes trekking ahead, doing the hard and dangerous work of breaking the trail.

Arctic saltwater freezes solid in level pans or floes that floated on the surface of the sea. These can be 12 feet (3.6 m) thick, often miles in extent. The continuously flowing ocean beneath them affects their size and shape. Floes can part to create widening lanes of open water known as leads, then rotate and smash into each other, driving up huge ridges of broken ice towering as high as 60 feet (18 m).

From the Arctic explorer's viewpoint, any path made over the floes would soon be lost. The constant shifting of ice as it broke up, divided or merged with other floes, would obliterate the location. This meant that anyone intending to travel any distance in this part of the Arctic had to keep all their necessary supplies – food, cooking fuel, and housing – with them at all times. There was no possibility of creating a depot of supplies and food to be found on a return route by retracing one's steps.

Unlike most of the Arctic explorers who had preceded them, Peary and Henson adopted the ways of the local Inuit for living in this forbidding environment. Rather than carry canvas tents, they built an igloo whenever shelter was needed. They abandoned traditional sleeping bags in favor of the Inuit fur clothing worn day and night. The weight saved allowed for an additional 16 pounds (7.2 kg) of pemmican (a fat- and protein-rich food) to be carried, giving their team an additional two days of travel to get them that much closer to the Pole.

They left solid land on April 6, 1902. During this first real foray onto the floating ice, the trail was brought to an early halt by a semi-permanent "river" of open water between ice floes that was given the name "Big Lead." It was far too wide to cross on the thin new ice that occasionally formed on its surface.

Peary sent everyone back except for Henson and two Inuit. The expedition paused to regroup at its southern edge, waiting an entire week for the river to close of its own accord. When it did on April 14, the expedition hurried over, with Henson and the Inuit breaking the trail ahead while Peary sat idle in camp.

On April 21, 1902, after 16 days of hard-fought advance, they had reached the limit of their supplies and their endurance. Peary claimed a new farthest north record, 84° 17' and established his method as a workable approach to reach the Pole. He also realized that he would need a new ship designed specifically for task, one that wouldn't be stopped or damaged by the shifting ice.

Chapter 10
Exploration

'North Pole'

Peary's design for his new ship SS Roosevelt followed the example set by Nansen's 1893 exploration ship Fram, with its hull rounded below the waterline. This hull design enabled the ship to absorb the impact from the drifting ice floes against the ship, and to rise up when the ice closed in around it, preventing it from being crushed.

The SS Roosevelt was named after the US President at the time, the determined and dynamic Theodore Roosevelt, who was an admirer of Peary.

Inside the ship, the massive framing made the Roosevelt's hold too full of beams and cross-bracing to carry everything needed for the expedition so, an auxiliary ship, the Erik, was commissioned to carry the extra coal the Roosevelt needed to reach into the upper part of Baffin Bay.

After the team offloaded the last supplies from the Erik and sent her home, the Roosevelt continued onward, stopping for the winter of 1905 at Cape Sheridan. The men and dogs then trekked another 90 miles (145 km) overland along the coast to Cape Columbia. Their sleds on that journey contained all the supplies needed for the march over sea ice to the Pole. Peary freely entrusted Henson with whole segments of the operation.

Once everything was in place at Cape Columbia, the expedition was ready to set out over the sea ice in its final push for the Pole. The changeable ice conditions dictated what happened next.

The floes drifted in seemingly random directions. One day's progress would be negated by the unseen, unfelt movement of the floe on the ocean current. At times the men, at the end of a hard day's travel, would be farther from their destination than when they started.

Only constant movement by the support parties would keep the short segments of the trail open, making them easier to find by the less experienced division leaders. The advance teams moved slowly until they were stopped altogether by the Big Lead. After eight days, the Lead froze over sufficiently to allow them to cross it.

Once on the denser pack ice the advancing teams could bypass each other; the second team using the igloos built by the first for their resting stops. Keeping the trail open for the return journey would enable them to reuse these station points (if the igloos were still standing), saving valuable time and energy.

Another farthest North: 87° 6' North, April 21, 1906

The Inuit sled drivers understood the conditions better than any of the Americans. Their goal was to survive and get home. They knew the season was becoming too advanced causing the ice to become too fragile or too jagged from pressure to make the necessary progress.

After a series of long drives, it finally became clear to Peary that the remaining distance to the Pole – well over 200 miles (322 km) – was far too great to achieve before the advancing spring season rendered it impassable.

Their return journey was impeded by another big lead, dubbed the "River Styx," that had opened during their absence. Crossing on ice so thin that the toes of their mukluks were breaking through to the cold water beneath, Peary, Henson, and the rest of the expedition team barely made it back alive. Of all the men that day, Matthew Henson deserved the greatest credit for directing the work of the Inuit drivers and guiding the additional American team members safely back to solid land.

The problem was that they did not arrive back to the shore they had left from. They found themselves much farther east (actually on the northern coast of Greenland). It was here that Peary first discovered how brutal the ocean currents beneath the ice could be. Any attempt to go north had to head northwest first, somewhat by dead reckoning, to counteract the eastward drift of the ice floes. Peary and Henson tried again for the Pole in their third season at Fort Conger, but their efforts produced no more "nearest-the-Pole" records.

Their record for farthest North, 87° 6' latitude, would have to suffice. Despite this heroic journey over the sea ice and back, the sting of Peary's latest failure would continue to rankle.

Peary was wise enough to realize that his initial idea of using minimal support teams to help him get to the Pole was not going to work. The only way to reach his goal was to ensure enough supplies could be brought over the shifting ice. It would have to be done in relays.

To achieve this, his next expedition would need to include larger teams of men. The entire operation would need more Americans and more Inuit dog-drivers to be brought to the most northern shore possible, using the SS Roosevelt once again.

One more try

Henson's ambition was closely tied to Peary's.

When Peary was ready to make another attempt, so was Henson. This is in stark contrast to many of the Americans (all of whom were white men) on the 1906 expedition who had been too overworked and disappointed by lack of success to strive for the North Pole again.

It was clear that even more Inuit under Henson's authority would be needed to ensure enough supplies were moved far enough out on the ice to keep a trail open for a rapid trek. A substantially larger party of white men were hired for the job and taught the basics of Peary's plan. While Peary was the leader of these expeditions, Henson – clearly now second-in-command – would manage the work of the American men as well as that of the Inuit.

After a send-off that included President Theodore Roosevelt joining them on board the ship to wish them farewell, Peary sailed the *Roosevelt* to the far north again, anchoring it at Cape Sheridan in September 1908. Henson immediately took charge of three Inuit men and set out on a hunting trip to bring back enough musk-oxen meat to feed the men and the dogs for the next six months.

As before, the plan was to have discreet teams moving supplies northward by dog sled in the autumn, leapfrogging each other as they traveled along the difficult route from Cape Sheridan to Cape Columbia.

This was no easy task.

On October 14 twelve sledges, each pulled by ten dogs, were lined up on the ice, ready for departure. The men and dogs were divided into five independent, self-contained divisions intended to move everything forward in the most efficient manner. Each had one white American in charge. Henson directed the overall work and acted as interpreter to the Inuit.

This labor of transferring the supplies from Cape Sheridan to Cape Columbia continued uninterrupted through the winter and early spring of the following year, which was by far the coldest time of any year. On February 26, 1909, the thermometer stood at -58° F (-50° C).

Floating ice pans in the Arctic. *(Photographer: Barbara Rae)*

The following night was even colder. The whiskey and gin were frozen solid, and their vital kerosene fuel for cooking had turned into a watery, white mush. David MacMillan, one of the division leaders, even noted that the Inuit declared it was the coldest night they had ever experienced.

The big push

On March 1 1909, the entire expedition – 24 men, 19 sledges, and 123 dogs, left Cape Columbia and set out across the sea ice. Peary would not be in the lead, but he did set the pace and establish the course. His missing toes rendered him unable to walk quickly or far, so whenever possible he traveled as a passenger on one of the sledges. Captain Bob Bartlett, Henson and others would go on north ahead of him, breaking the trail and leapfrogging past each other, setting up igloo camps along the way.

The course was set a little west of true north to counteract the underlying sea current from west to east, previously discovered by Peary.

Henson and the white men took turns in the lead, pushing forward as far and as quickly as the shifting ice pans would allow. With no fixed ground beneath him, Peary navigated by sextant for sun sights to determine latitude, and magnetic compass which, this far North, actually pointed toward the southeast. Because of this, their actual position on the globe was largely educated guesswork.

This attempt on the Pole had every appearance of success. Years of practice in the complex scheme of alternating leapfrogging advance parties, and an early stretch of good luck in weather and ice conditions enabled the expedition to get off to an excellent start. But similar to earlier expeditions, they were held up for a week waiting for the Big Lead to close. Once the gap closed, the expedition made good mileage – up to 15 miles (24 km) a day. Given their speed when ice conditions allowed, they seemed to have sufficient time to reach the North Pole and return from it safely before the spring weather weakened the ice.

That good fortune was about to end.

The floating ice was causing pressure ridges 50 feet (15 m) tall. These obstacles to sledging had to be broken through or surmounted at great effort, sometimes resulting in the capsizing of the sledges. When the sledges became damaged on the trail, Henson was the man to repair them. When they were broken beyond repair, he'd make a new one from their remains.

Peary's plan was to have four supporting parties in addition to his own Pole party to keep the trail open and visible, so that a constant flow of support teams could pass and bypass each other. But not all the teams were equal in strength or capacity. The weaker ones, those with worn-out and crippled dogs, turned back with half of the food they would need to reach Cape Columbia. They would have to double their outward marches, but it was doable since they were no longer burdened with supplies intended for the final Pole party.

Captain Bob Bartlett and his Inuit dog drivers were the last supporting team to leave Peary and Henson. Bartlett, the only other competent navigator among the expedition team to make it this far North, calculated their position as 87°40' on April 1, 1909. This was the last accurate measurement to be taken and recorded.

The men who went the farthest

After Bartlett's team turned back, only Peary, Henson, and the four most capable Inuit men – Egingwah, Ooqueah, Ootah, and Seegloo – remained to go onward.

The obstacles and dangers were unremitting.

On April 3, Henson was driving the last sledge in the line, when his foot slipped on a block of ice at the edge of the ice flow. In an instant, he was thrown into the freezing water, unable to grip the ice-edge with his gloved hand. "Before I knew it," he wrote later, "faithful old Ootah grabbed me by the nape of the neck, the same as he would have grabbed a dog, and with one hand he pulled me out of the water, and with the other hurried the team across." Ootah's quick action had saved his life.

After several days of further marches, on April 6, Peary called a halt. The temperature was -29°F (-34°C). This was to be their furthest north camp. The final victory of Peary's personal goal.

He unfurled the silk American flag made by his wife Josephine, the same flag he had brought on all his and Henson's expeditions and flown at all their last camps. Peary fastened the flag to a staff and declared:

"We will plant the stars and stripe – *at the North Pole!*"

The Inuit drivers gathered round and gave three hearty cheers.

As the sound of their voices died away, Henson felt a thrill of patriotism as the flag rippled in the frigid breeze. "I felt all that it was possible for me to feel. That it was I, a lowly member of my race, who had been chosen by fate to represent it, at this, almost the last of the world's great work."

Peary left him there to establish their encampment and, taking two of the Inuit, went on ahead to take further sextant observations to confirm that they really had reached the North Pole of the Earth.

To Henson it was a moment of glory. It would be Henson's own personal best, his farthest north, an achievement not for reaching a destination, but for having invested the vast resources of his own time, energy, skill and perseverance to get there, for himself and for Peary.

At the time, Henson was unaware they had not actually reached the North Pole. But Peary knew it. The moment of their farthest north, *near* but not *at* the North Pole, was the scene of Peary's greatest folly and the downfall of his life's dream.

When Henson stepped forward to shake his boss's hand in sincere congratulation, his outstretched, ungloved hand was ignored. Peary passed him by and crawled into his own igloo.

Other than the briefest of commands and orders necessary to get the polar team safely back home, Peary never spoke to Henson again.

Henson spoke candidly about that moment in an interview published in the *Boston American*, in July 1910. "From the moment I declared to Commander Peary that I believed we stood upon the Pole, he apparently ceased to be my friend… After twenty-two years of service with Peary we are now as strangers."

This is where their relationship that sustained their mutual endeavors over the years came to an abrupt end.

For Peary, perhaps they had never been a team. For Henson, he believed the opposite. Their aspirations may not have been exactly the same, but their passions were. Their combined efforts over the years had brought results that could never have otherwise been achieved. Was that not sufficient for a moment of shared congratulations?

Regardless of how those two men viewed their relationship from within, there is no question that they were a team – one of the most enduring, intriguing and accomplished polar teams that ever existed. Neither could have achieved their farthest north results without a deep understanding of each other and their expectations. Both did their utmost to reach a single, shared goal.

And whether they actually stood for a moment on the axis of the earth's rotation, their remarkable achievements cannot be denied. Together, they had pioneered new routes and improved methods of polar travel; mapped previously unseen land, locations and coastlines; determined that Greenland was a large island and not a continent that extended to the North Pole; explored remote parts of the Arctic; and persevered through numerous obstacles and setbacks.

We can judge Peary and Henson by modern standards or we can recognize that their relationship was shaped by the times in which they lived. Either way, they broke down race barriers, and Josephine's participation in the Arctic eliminated gender barriers for all future, adventurous and courageous women seeking to participate in polar expeditions.

Did they reach it?

Peary and Henson returned to civilization in 1909 to find that another American, Dr. Frederick Cook (who had accompanied Peary, Henson and Josephine on their first Greenland expedition), claimed to have launched a different expedition and had reached the North Pole in 1908, a full year before Peary and Henson. Cook's return had been so difficult that it took him a year to get back to civilization.

The news of their competing claims took the world by storm. Two accomplished American explorers, Cook and Peary each claimed that he and his team had reached the North Pole, and the other had not. Neither had kept sufficient, definitive records to confirm their accomplishment.

The reality was neither had reached it.

The teamwork developed between Henson and Peary over their 22 years of working and exploring – in hardship and in triumph – ended on that final day, nearest the North Pole.

Peary seldom spoke of his erstwhile teammate after his final Pole expedition. Henson lived much longer, and occasionally expressed his memories and feelings in articles and interviews, and in his 1912 autobiography *A Negro Explorer at the North Pole*. Peary's introduction to the book, written at the behest of the publisher, does not square with his treatment of Henson at the Pole. Henson remained proud of their accomplishments.

Henson died in 1955, having outlived Peary by 35 years. Years later, Henson's remains were moved to Arlington National Cemetery and buried alongside those of Peary.

The joint efforts of these two men have helped to complete our understanding of the amazing planet on which we live. The way Greenland appears on any globe or world map today is directly attributable to their expeditions.

The first person to actually reach the North Pole by dog sledge, the way Peary and Henson tried to do, was Wally Herbert in 1969 – the same year another two-person team, Neil Armstrong and Buzz Aldrin, stepped foot on the moon.

★

There was another very small team that came together to solve a specific problem. They worked at the opposite end of the world, Antarctica. Their endeavor was to prove a Victorian theory of evolution that involved penguin eggs.

Theirs may have been the worst journey in the world, but it is the best showcase of how a team of three can overcome the harshest conditions imaginable.

At the start of the Winter Journey (*from the left*) Bowers, Wilson, Cherry-Garrard.
(Photographer: Herbert Ponting, 1911)

Chapter 11
Science And Discovery

Wilson, Bowers and Cherry-Garrard

"We did not forget the Please and Thank you, which mean
much in such circumstances, and all the little links with
decent civilization which we could still keep going."

—*Apsley Cherry-Garrard*

Few have suffered so greatly and for so little reward than the three intrepid explorers who left the comfort and safety of their expedition hut for a dangerous, and possibly, foolhardy six-week overland journey during the deep-black heart of the Antarctic winter of 1911.

These men bonded during the long, risky, and troubled voyage at the start of the British Antarctic Expedition 1910-1913 led by Captain Robert Falcon Scott. Scott had previously captained the British National Antarctic Expedition 1901-1904 which was also known as the *Discovery* Expedition. Now the time had come to build upon the scientific successes of that earlier trip.

Their ship, the *Terra Nova*, dropped the three men and the other twenty-four members of the shore party at the frozen beach at Cape Evans, on Ross Island, Antarctica, in January 1911. This was the largest and best-equipped scientific and geographical expedition yet to land at the frozen continent.

The *Terra Nova* Expedition's plan

There were unexplored territories in every direction, from barren tracts of ice and snow to mountain ranges and hidden valleys – all of them brimming with geological mysteries. The expedition would measure and record in exacting detail glacier movements, weather patterns, magnetic variations, the pull of gravity, and the flora and fauna of the region. The team couldn't have known it at the time, but all this data would eventually become the baseline used for these Antarctic sciences today.

Captain Scott had an additional aim: to be the first to the South Pole.

Scott's plan was for living quarters to be erected by the men immediately upon arrival at Cape Evans. The scientists and explorers would live in relative comfort ashore in a spacious (for the conditions) prefabricated hut – 25 feet by 50 feet (7.6 by 15.2 m), double-walled and insulated, and heated as best as it could be against the bitter cold. It would be packed with foodstuffs and scientific equipment of every discipline.

Just outside the walls of the hut would be unheated storerooms built from packing crates, and an enclosed stable with bales of fodder to house the ponies. These animals would be used to pull sledges filled with supplies. There would be sledge dogs as well, and a few experimental motor tractors.

Scott's South Pole strategy was that starting in October, when the sun will have risen above the horizon after the long, dark Antarctic winter, a portion of the team would embark on a hazardous 800-mile journey (1,287 km) over many months. Their goal would be to trek across the ice and snow to reach the South Pole, claiming for their homeland this great geographical prize.

This would be a daunting endeavor. Scott and his men would follow a route pioneered by Scott, Edward Wilson and Ernest Shackleton in 1902, and then pushed further South by Shackleton and Frank Wild, who almost lost their lives in the process in 1909.

Shackleton and Wild's adventures are described later in this book.

The first part of the route Scott and his men would traverse could be aided by using ponies, dogs and motor sledges to help pull the sledges. But once they reached the mighty and crevasse-ridden Beardmore Glacier, the bulk of the journey would be done by the men, donning harnesses and pulling the sledges themselves. That trek, if successful, would take them onto the South Polar plateau, achieving elevations of 10,000 feet (3,048 m) above sea level. It would be several hundred miles along that plateau where they would find the South Pole.

To aid in this journey, Scott's plan was that once the hut at Cape Evans was constructed, men and ponies would set out over the ice, pulling sledges laden with field supplies including food and cooking oil. This would be offloaded and set up as a depot as far south as they could get, so the South Polar team could retrieve items from it on their outbound and return journey.

The men tasked with depot laying finally set off, trekking south from Cape Evans. At 79° 29' South latitude, which was 130 miles (209 km) from the hut, they called a halt, and unloaded the food and fuel. They built a cairn of snow blocks over the stockpile, marked the place with flags, and named it "One Ton Depot" based on their estimated weight of supplies there.

They then headed back to Cape Evans.

Birdie, Cherry and Uncle Bill

Three of the men on that depot-laying journey had already become fast friends.

For "Birdie," "Cherry," and "Uncle Bill," their common cause was a scientific venture that was still aligned with the goals of Scott's expedition: to fully explore the Antarctic continent within their reach. Shortly after the shore party's midwinter celebration, these three men prepared to make a 35 mile (56 km) stretch of it their very own, in ways they could barely have imagined.

They hadn't known each other long, but – along with everyone else on the *Terra Nova* – time spent at sea and coping with storms that could have sunk the ship can turn total strangers into lifelong friends. In the close quarters of a sailing steam whaler the men on board would size each other up quickly. In the throes of a near disaster, each man shows his truest mettle. There is he who panics when the ship seems about to sink – as nearly happened en route to Antarctica – and there is he who bucks up and pitches in with a smile on his face.

"Uncle Bill" was the nickname given to Dr. Edward Wilson, the expedition's Chief of Scientific Staff, but he was much more than that. He already had three years of Antarctic experience under his belt, more than almost anyone in 1910. Aboard Scott's *Discovery* Expedition, Wilson had been Assistant Medical Officer, Artist and Vertebrate Zoologist. A close friend and confidante of Captain Scott, he extended his measured wisdom to everyone else on the expedition as well.

"He is the soundest man we have," wrote Birdie in a letter home, "a chap whom I would trust with anything." Wilson returned the compliment in one of his own letters home: "He [Birdie Bowers] is a perfect marvel of efficiency – but in addition to this he has the most unselfish character I have ever seen in a man anywhere."

After setting sail from Cardiff, England on June 15, 1910, Apsley Cherry-Garrard began taking note of his shipmates: "Bowers was proving himself the best seaman on board, and with a supreme contempt for heat or cold."

These first impressions would hold true through all their shared tribulations in the coming months.

With no sea or polar experience of his own, Apsley Cherry-Garrard had joined the crew at Cardiff, paying his way on board with a sizable contribution of £1,000 (approximately $150,000 in today's money) to take the role of "adaptable helper." Cherry-Garrard had attended Winchester College, an elite high school that focused more on building character, toughness and physical and mental resilience in its students than teaching academic subjects.[10]

---

[10] Winchester College in the days of Cherry-Garrard and George Mallory sounds similar to

He was at Winchester College at the same time as George Mallory, one of the greatest British mountain climbers who ever lived. Mallory died on Mt. Everest as part of another very small team, one we'll be looking at later in this book.

Over the course of the expedition, Cherry-Garrard's adaptability became ever more apparent. One of the many skills he acquired was that of unofficial expedition historian, a role that led him to write his much-admired book *The Worst Journey in the World*. That title refers not to Scott's tragic return from the South Pole in 1912, but the memorable Winter Journey that Cherry-Garrard would soon undertake with Dr. Wilson and the estimable "Birdie" Bowers.

Lieutenant Henry Robertson Bowers, better known as Birdie, had been a member of the Royal Indian Marine Service when he first learned of the upcoming expedition. He made up his mind that he must be a part of it.

Bowers was every bit as adaptable as Cherry-Garrard. He immediately found his role as overseer of the loading of gear and supplies into the holds and onto the decks of the *Terra Nova*. Last to come aboard were the 34 dogs, 19 ponies, and the three innovative motor-tractors that Scott hoped would help speed the expedition partway to the Pole, taking them over the level sea ice and snow, at least up to the Beardmore Glacier.

Places had to be found for every item their shore party would need for the next two years, including the frame for the hut. So much was packed onboard that the 187 feet (57 m) *Terra Nova* was seriously overloaded. The Plimsoll mark on the hull's side, intended as a measure for maximum safe loading, was under water. The ship – a Scottish whaler built in 1884 – was no longer new, and her long passage through the tropical heat of the equator had dried out her decking and opened seams between the planks. Less of a problem in calm weather, but when the ship was hit by a huge storm on the way South, she came very near to foundering, never to be heard from again.

Gordonstoun School in Scotland, where King Charles was sent as a young boy, as portrayed in the television series, *The Crown*.

The *Terra Nova* and the expedition barely survived. It was during these terrifying days that the friendship between Wilson, Cherry-Garrard, and Bowers began to form.

And it was during the harrowing weeks of their Winter Journey that they formed a bond only death could sever.

The plan

Wilson's goal was to recover unhatched emperor penguin eggs to prove a theory of evolutionary science. The prevailing scientific idea at the time was that emperor penguins were primitive birds closely related to the reptilian family, and that a close look at the early forming embryo in a recently laid egg could prove Darwin's Theory of Evolution linking reptilian dinosaurs to birds.

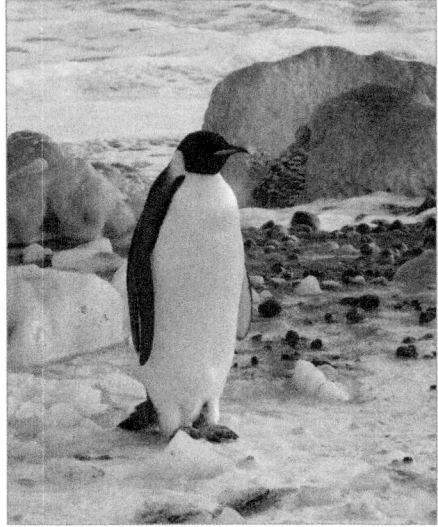

Emperor penguin in Anarctica.
(Photographer: Helen Wernham)

The challenge, however, was that emperor penguins laid their eggs during the heart of the Antarctic winter.

Wilson discovered this fact during Scott's *Discovery* Expedition. His plan now was to take the two best men he knew – Birdie Bowers and Apsley Cherry-Garrard – on a grueling midwinter journey to the emperor penguin breeding ground at Cape Crozier, 35 miles (56 km) from Cape Evans. This would be at a time when an Antarctic "day" was 22 to 24 hours of darkness.

Dragging six weeks of supplies on two sledges, the men planned to build a stone "igloo" on arrival at Cape Crozier. They would use it as their base while venturing down over the crevassed shore ice to the floating sea ice, where the emperors were laying and nurturing their eggs.

The intrepid explorers would then seize a few eggs and, keeping them warm in their mitts, bring them back to the relative warmth and safety of their stone igloo. Wilson would then dissect the penguin embryos and pickle them for closer investigation back at Cape Evans, and later in the laboratories of the Natural History Museum in London.

That was the plan. What happened next is hard to appreciate as we read this, comfortable in our warm houses.

Had they known the risks involved in this quixotic pursuit, would they have set out on this scientific venture?

Their equipment was deficient in almost every imaginable way. Their tent, rations, sledging gear, clothing, and sleeping bags were all inadequate to survive temperatures that would fall to -75.8° F (-60° C). And that was without adding wind chill to the measurement. These temperatures were colder than any human had ever experienced.

Their method of travel was manhauling – not using dogs or ponies to pull the sledges, but strapping on sledge harnesses and pulling the heavily laden sledges themselves.

Everything they depended on failed them.

Only through personal stamina, a shared integrity, and devotion to their cause and to each other did they make it through their travails to triumph over some of the most daunting extremes ever faced by men, before or since.

Two could not have done it. A different three could not have done it.

These three barely did.

What they achieved is a testament to the human spirit and team camaraderie. Their story, told in the next two chapters, is unforgettable. Its outcome is equally shocking.

Chapter 12
Science And Discovery

A Journey Like No Other

Had Wilson, Bowers and Cherry-Garrard known on departure the trouble they were about to encounter, they might have had second thoughts about the wisdom of their journey.

Even at noontime on a cloudless day, it was near pitch black, with only the white snow and ice underfoot glowing ghostly to guide them in the dim starlight. Their journey had been timed so that their arrival at Cape Crozier would coincide with the light of a full moon. They had planned to use that light to pick their way down the slopes of Mt. Terror and onto the sea ice below.

Wilson, Bowers and Cherry-Garrard had company at first. Three of their colleagues from Scott's expedition team pulled the two sledges for them, relieving that burden at least as far as the Ice Tongue seven miles (11.2 km) south of the base camp hut at Cape Evans. Three more tagged along as well, jolly company for at least the first few miles, until the beastly cold on their faces told them it was time to go home.

From there, it was *only* another 35 miles (56 km) to their destination at Cape Crozier on the eastern end of the island. Fit men should make it in three or four days, under ordinary conditions. A few days then at the Cape to build their stone igloo, a week or so collecting specimen penguin eggs, another week homeward bound. Although provisioned for five weeks, they were expected home by the second week of August.

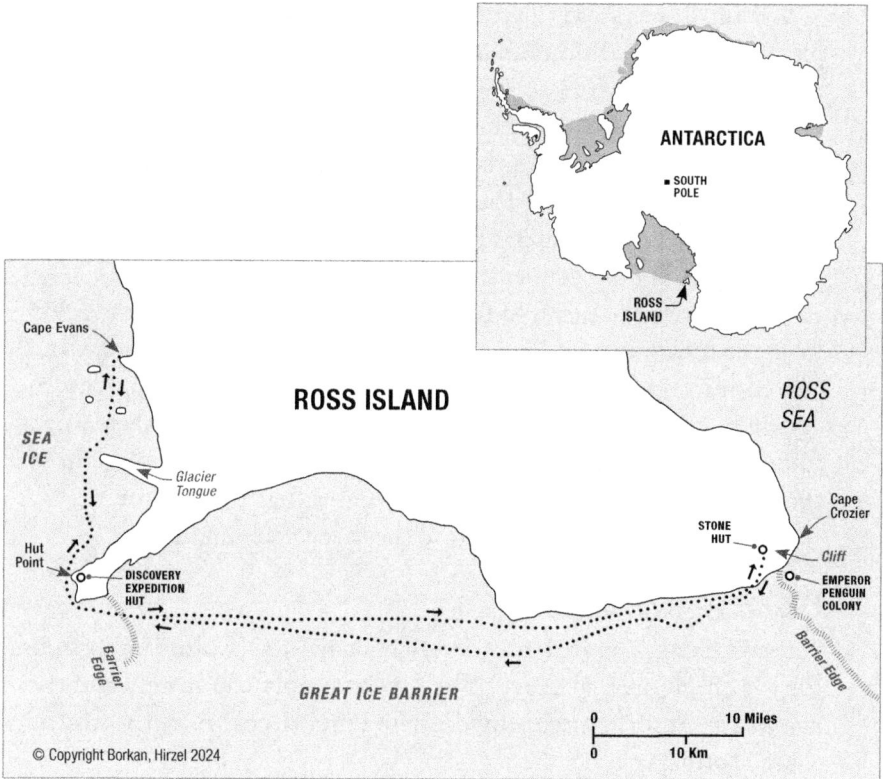

Winter Journey route based on a map drawn by Edward Wilson.
(*Locations and distances are approximate.*)

Once at the Ice Tongue, their helpers bid them adieu. Now Wilson, Bowers and Cherry-Garrard were alone with all their gear loaded onto two sledges, and a long walk ahead of them. The sledges were strangely difficult to pull. The snow beneath the runners acted like fine, white sand, creating more grip than glide.

Their slow progress meant they would not reach the last outward-bound way station – the old Discovery Hut – that night. Wilson knew it well since it had been erected during Scott's first Antarctic expedition. It was 12 miles (19.3 km) out from their start at Cape Evans. Instead they would get their first taste of the rigors of winter camping.

Wilson, Bowers and Cherry-Garrard already had plenty of practice pitching the expedition's pyramid tents, but never in the dark like this and never in cold so deep, exposed fingers would freeze in a matter of seconds. Despite these hindrances, they managed the task. Brushing the snow off their jackets, they climbed through the entry door and, shedding only their boots, wormed their way into the frozen reindeer-fur sleeping bags.

Sleep would only come after hours of violent shivering.

Wilson's wake-up call in the "morning" brought welcome relief from the incessant cold, a chance to have a bit of biscuit or pemmican stirred into their breakfast tea, the promise of physical activity to course hot blood into their frozen extremities. They quickly learned to bend into their sledging postures immediately on coming out of the tent, as their clothing would freeze iron-hard in a moment.

The men would have to endure the same each night.

For every step forward

The two sledges, pulled in a train-like formation, became impossible for the three men to move. They were simply too heavy, and the snow was unforgiving, providing no smoothness underneath the sledge runners.

It became clear to Wilson, Bowers and Cherry-Garrard that they would have to move the sledges in relays, dragging one forward a mile or two, then returning for the other, to bring it forward as well. Wilson's diary is full of laconically pencilled details like: "The temperature remained at -50° F. all day, and we felt the cold a good deal."; "We made only 3-1/4 miles today, but walked about 10."; "We made in all our 2-1/4 miles today – very slow work."; "Our sleeping bags are beginning to get wet thanks to these low temperatures."; "We relayed for 8 hours and only advanced 1-1/2 miles for the day."

The midwinter temperatures then dropped to potentially lethal lows.

At -76.8°F (-60.4°C), they were experiencing the lowest temperatures ever survived by anyone on the planet. Their clothing grew ever heavier, ever more iron-like, as their sweat froze into it. It took as long as two hours to thaw a frozen sleeping bag with body heat and work one's way into it. But as Dr. Wilson knew, they must have their seven hours' rest, even if the violent shivering kept them awake all night.

In his iconic book about this journey, Cherry-Garrard explained that even at night their clothes were cold with damp and parts were ice-encrusted. Their fur sleeping bags were so frozen and iced over that they felt like sheets of armor.

Wilson asked Bowers and Cherry-Garrard if they thought it might be better to give up, to turn around and save his penguin embryology for another expedition. Each time he asked in his calm, unruffled manner, "Shall we go on?"

And each time Bowers and Cherry-Garrard would reply: yes, the hardships were worth the goal. No other man but Wilson could have led them this far. At all times Wilson displayed exceptional leadership that resulted in astounding levels of team camaraderie.

So they kept going on, sometimes covering only a mile or two for all their effort, in dogged pursuit of one of science's nobler goals.

Chapter 13
Science And Discovery

The Windsept Terrace

After nineteen punishing days, they came upon a windswept terrace overlooking the emperor penguin rookery at Cape Crozier. Here they built a little stone hut, their "igloo" of four low-stone walls. Its roof was a single layer of green canvas cloth attached to upright ski poles.

Within this sparce shelter they would stay for a few days, bringing back new eggs warm in their gloves for Wilson to dissect and preserve in search of his elusive evolutionary evidence. They would also bring eggs to eat, as well as fresh penguin meat to cook over an improvised blubber stove.

That was the plan. The very lap of luxury after their ordeal.

In the semi-darkness of midday, through much toil and danger, they were at last able to get down to the penguin rookery and successfully retrieve five eggs from those remaining.

They placed the eggs intact in their gloves, and also carried three penguin carcasses to use for food and blubber. Cherry-Garrard suffered from poor eyesight and couldn't wear his metal-framed spectacles in the extreme cold, so with his vision limited, the two eggs he was carrying broke.

Upon reaching their stone hut, the three remaining eggs were placed in alcohol.

In building their stone hut, they had not considered why the gravel summit of the rise was bare of snow. They soon learned first-hand about the furious gales that swept up off the Barrier that kept it clear.

The wind rose to a hurricane level then to a full-on blizzard. One furious blast ripped the green canvas roof to shreds. The snow poured into the roofless hut, covering them deeply as they lay in their bags praying for an end to the gale. They were exposed to the full ferocity of the icy tumult. At the same time, the fierce wind also tore away their only tent. It had been pitched outside protecting their food crates and sledging supplies.

Captain Scott recorded the tale as he must have been told by either Wilson or Cherry-Garrard upon their return: "Bowers put his head out once and said, 'We're all right,' in as near his ordinary tones as he could compass.' The others replied, 'Yes, we're all right,' and all were silent for a night and half a day whilst the wind howled on."

Bowers noted the event in his diary with his usual aplomb. "I was resolved to keep warm and paddled my feet about and sang all the songs and hymns I knew to pass the time. I could occasionally thump Bill and as he still moved I knew he was alive all right. What a birthday for him!" It was Wilson's 39th birthday.

Few men have ever been so destitute, so far from succor as Edward Bill Wilson, Birdie Bowers, and Apsley Cherry-Garrard.

With no tent, they could not hope to survive the long journey home to the hut at Cape Evans. Ordinary men might have succumbed to despair. But these three had an indomitable sense of teamwork to sustain them, despite the looming disaster.

Wilson wrote in his diary that night, "We were not really so much disturbed as we might have thought, and we had time (while the blizzard was raging over their heads!) to think out a plan for getting home without our tent—in case we couldn't find it—and without the canvas roof of the hut which had gone down wind in shreds the size of a pocket handkerchief."

Yet no one voiced a word of despair, a yielding to inevitable defeat. This is no exaggeration. Their three personal accounts – Wilson's in his diary, Bowers' in his letters home, Cherry-Garrard's book published years afterward as *The Worst Journey in the World* – all agree on this point. They had come to do a job, they had come this far to do it, and they were not about to give in to despair with so much yet to do.

The tent, by some miracle, had not been blown out over the frozen sea, but came to earth a half mile away. They would need it. Survival on the return journey depended upon it.

With only one tin of cooking fuel remaining for the entire return journey, it was time to leave. Wilson, Bowers and Cherry-Garrard could not risk spending time to retrieve more eggs. Their grandiose plans, the unparalleled misery of their outbound trek, the dreadful days in the ruin of their stone igloo never knowing if they would ever see home again, and other challenges too numerous and terrible to recount were behind them.

And so they started for home. "The journey home from here was by far the coldest experience I have ever had," wrote Wilson.

There was more daylight now. The few hours of twilight grew slightly longer each day as the sun rose a little bit higher even though

Photo taken immediately upon their return. *(from the left)* Wilson, Bowers and Cherry-Garrard. *(Photographer: Herbert Ponting, 1911)*

it still remained below the horizon. But the cold was no less brutal.

The return journey, although still bitterly cold, would go much faster thanks to the greatly reduced loads on their one returning sledge. One week later, the three weary men staggered into the warmth and safety of the hut at Cape Evans on August 1, 1911, after an absence of five weeks.

It was about the journey

Cherry-Garrard was the only one of the three to survive the next season's work in Antarctica.

Wilson and Bowers perished with Captain Scott on their return from the South Pole, after coming second to the Norwegian explorer Roald Amundsen. Scott, Wilson and Bowers died 11 miles (17.7 km) from One Ton Depot – the depot of food and supplies that Cherry-Garrard, Wilson, Bowers and others had established the year before.

Cherry-Garrard was part of the search party who looked for the remains of the Polar Party. Discovering his beloved deceased companions impacted him for the rest of his long life.

Cherry-Garrard wrote: "These two men went through the Winter Journey and lived: later they went through the Polar Journey and died. They were gold, pure, shining, unalloyed. Words cannot express how good their companionship was.

"Through all these days, and those which were to follow, the worst I suppose in their dark severity that men have ever come through alive, no single hasty or angry word passed their lips. When, later, we were sure, so far as we can be sure of anything, that we must die, they were cheerful, and so far as I can judge their songs and cheery words were quite unforced. Nor were they ever flurried, though always as quick as the conditions would allow in moments of emergency. It is hard that often such men must go first when others far less worthy remain."

As for the eggs

Upon his return to Britain, Cherry-Garrard brought the eggs to the Natural History Museum in London. The museum saw little of value in the donation and Cherry-Garrard had to wait most of a day just to receive a receipt for them.

When the eggs were later dissected and analyzed in the best laboratories in the country, the thawed remains of the embryos inside revealed thoroughly modern birds. The missing link that Wilson had sought – the evolutionary link between reptilian dinosaurs and primitive birds like the emperor penguin – was not to be found.

In the end, the Winter Journey was more about the three men who undertook it than about the eggs they brought home.

Edward Wilson, Birdie Bowers, and Apsley Cherry-Garrard shared an experience such as few very small teams can ever know. They had sacrificed equally in pursuit of Wilson's desire to prove (or disprove) the theory of penguin and dinosaur evolution – a goal of his that had become their own. They could have given up, but they never did.

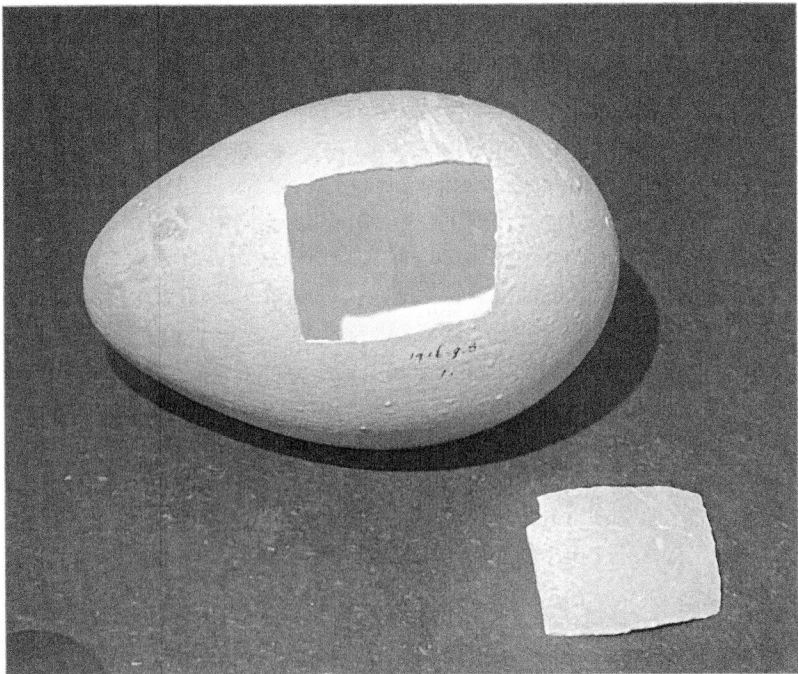

One of the three Emperor penguin eggs retrieved by Wilson, Bowers and Cherry-Garrard. *(Photographer: Brad Borkan)*

Cherry-Garrard wrote: "And we *did* stick it. How good the memories of those days are. With jokes about Birdie's picture hat; with songs we remembered off the gramophone; with ready words of sympathy for frost-bitten feet; with generous smiles for poor jests; with suggestions of happy beds to come. We did not forget the Please and Thank you, which mean much in such circumstances, and all the little links with decent civilization which we could still keep going. I'll swear there was still a grace about us when we staggered in. And we kept our tempers – even with God."

A team

These men had maintained a cocoon of solid friendships and carried deep respect to sustain them throughout their hardship. It was left to Cherry-Garrard to tell their story.

It is a story of how three people of different strengths and backgrounds came together and shared not only their energy but their aspirations. They saw a project through the most demanding of challenges, and accepted the conclusion, whatever it may have been.

To this day, the three penguin eggs are still in the Natural History Museum in London.

Two are kept in a remote location; one is on public display in their Treasures Room, for all the world to see. It sits among twenty-two of the most important objects related to the scientific natural world, including an original printing of Darwin's *On the Origin of Species*.

★

Now we turn our attention to very small teams who engaged in creative collaborations. They brought exuberance and energy to their work that changed how we appreciate and understand the cultural world around us.

Part 2

Creative
Collaboration

★

Eames House, Case Study House #8, Pacific Palisades, California.
(Photographer: Carol M. Highsmith) (Library of Congress)

Chapter 14
Creative Arts

Art Nouveau and Modernism

"Genius? Nothing – we just worked harder."

—*Charles Eames*

You might have thought, as we once did, that creating art would be a solitary endeavor. After all, only one person holds the paint brush or pencil, or presses the button controlling the camera shutter. We were surprised to discover a wide variety of pioneering pairs whose work in the fields of decorative arts, design, architecture, photography, and other media brought notice and acclaim not to an individual but to a team. And they did it with energy and creativity beyond what each individual could have done alone.

We initially considered art and design to be one of the "softer" endeavors, at least when contrasted with those of exploration, women's rights and designing airplanes. Yet, a closer look revealed it to be one of the most intriguing areas in which pairs of talented people teamed up.

While some were husband and wife teams, others had different relationships. The best and most proficient of these teams worked toward common goals, extending the boundaries of what art and design could be. Their creative legacy influenced the artists of their day, as well as those who came along years later. All of these artists in their own way influenced what the rest of us think of when envisioning compelling art and design.

Here are two examples of those teams.

Mackintosh and Macdonald: Art Nouveau

At the end of the 19th century a husband-and-wife design team emerged, not in Paris or New York as you might expect, but in Scotland. Charles Rennie Mackintosh and Margaret Macdonald met at the Glasgow School of Art in 1892 and later went on to create an Art Nouveau style that has become revered in the UK and throughout the world.

Mackintosh's early career in architecture had a boost in 1897, when the firm he was working for won the commission to design the Glasgow School of Art. The first part was completed by 1899. The designs featured large rectilinear framing for the huge sheets of glass forming the windows. It was dramatically different from contemporary architecture and set the stage for 20th century Modernism.

Working in a bygone era when women were considered second-class citizens (women in Scotland did not achieve voting rights until 1928), Charles and Margaret enjoyed a surprisingly modern marriage. They married in 1900 and were equal partners and collaborators.

They combined their talents to create new visions of interior and exterior architectural and

Art nouveau washstand designed by Charles Rennie Mackintosh. *(Metropolitan Museum of Art)*

furniture design. In the realm of architecture, their Art Nouveau work modernized design by weaving together geometric and structural shapes with elegant, flowing forms. Mackintosh and Macdonald's work became even more popular in parts of Europe than it was in Scotland – their home country.

Together they created beautiful, original designs in furniture, architecture, and interiors that were far ahead of their time and unexpectedly divergent from the conservative ideas of the day. Charles gave the credit for their successes to his wife.

"Remember," he once wrote, "you are half if not three-quarters of all my architectural talents." He went on to say, "Margaret has genius, I have only talent." Together they designed a number of key buildings and interiors, including Miss Cranston's Tea Rooms and the Willow Tea Rooms, both in Glasgow. A second section of the Glasgow School of Art was designed and built between 1907 and 1909.

As architectural fashion began to change, Art Nouveau evolved into Art Deco, and by the advent of World War I, commissions for their work dried up. Charles and Margaret moved from Scotland to the eastern coast of England.

Trouble followed.

Charles walked with a limp and suffered from a drooping right eye. He also had a thick Glaswegian accent. He was accused by the British locals of being a spy for Germany. His crime? Corresponding with artists and others in Vienna, and enjoying evening walks along the coast, guided by the light of a lantern. He was arrested and falsely imprisoned, under the accusation of signaling enemy ships with the lantern. Margaret secured his release.

Financially, the pair never recovered from the lean war years and they died nearly penniless. Yet, their enormous legacy lives on in the flowing Art Nouveau movement in art and architecture, and the succeeding industrial design of Art Deco. These two styles mark the first two decades of the twentieth century, and many great architects and designers who came after have studied the seminal work of the Mackintosh-Macdonald team.

In the UK and beyond, Mackintosh and Macdonald are renowned for innovative decorative design and considered to be one of the founders of modernism in architecture, furniture, and industrial design. Their buildings are visited by people from around the world, and with the enthusiastic revival of Art Nouveau in the last 60 years, their work today is highly prized by collectors and enthusiasts.

In 2014, the Glasgow School of Art, designed by Mackintosh (surviving drawings are unmistakeably his), suffered a disastrous fire. The fire department acted promptly saving 70 percent of the interior and 90 percent of the overall structure, but the beautiful library – often cited as one of the best examples of Art Nouveau architecture and design in the world – incurred considerable damage. The actor Brad Pitt and other celebrities helped raise tens of millions of British pounds to rebuild it.

With restoration and repairs nearly complete, in 2018 another devastating fire occurred. Again, the building is being lovingly restored, proof of how influential Macdonald and Mackintosh were on this important art form.

Charles and Ray Eames: Modernism

Another example of a very small team extending an art form was that of Charles and Ray Eames, who started collaborating in the early 1940s while World War II was still underway. Their work became even more important in the immediate aftermath of that war.

The late 1940s and early 1950s was a unique period in the United States. With returning soldiers, a growing middle class, and the start of a baby and housing boom, there was a burgeoning requirement for modern houses. Consumers desired stylish furniture and home decorations that were functional, durable and affordable.

Europe and Asia were still recovering from the effects of the war and could not contribute to America's growing needs effectively. For American designers, materials had to be sourced from the USA. To meet market demands, speed of design and manufacture was of the essence.

Charles Eames was the all-American boy. Athletic, smart,

handsome, and popular in high school. He then studied architecture, and turned a fellowship at the Cranbrook Academy of Art into a paid Industrial Design department position. His first marriage faltered, and while separated from his wife, as the newly appointed head of the Industrial Design department met a female student named Ray Kaiser who was studying in a different discipline.

Charles and Ray dated, married, and moved to California. They worked together harmoniously for the rest of their lives.

They are probably best known for the Eames Lounge Chair and Ottoman, which looks surprisingly modern, even today. The number of objects they designed was enormous and varied, ranging from house designs to home furnishings. They also created over one hundred films, and over two million photographs, artworks, aircraft parts, home accessories, toys, exhibitions, lectures, slide shows, and corporate branding projects. They pioneered 3D plywood molding techniques and fabricating chairs using fiberglass, wire mesh and other materials.

Although Ray designed stylish magazine covers alone for *Art & Architecture* magazine and they each designed pieces for a Museum of Modern Art competition, almost everything else Ray and Charles did together. They designed modular office furniture and the familiar tandem seating for airports consisting of four or five seats bolted to a single frame with armrests at each end. Modern office furniture and airport seating in use today not only were all influenced by the Eames work from the late 1940's through to the 1980s, but some of it is still Eames furniture. They intended for the furniture they designed to be mended and repaired and passed on to future generations.

Their desire to work with low cost, fast-to-manufacture materials led to them extensively working with plywood. Plywood is made of thin sheets of wood glued together. The glue in use at the time was not strong enough to hold during the steam-heat process used to bend the wood into molds for their furniture designs. Charles and Ray solved this problem by inventing new methods to make plywood, experimenting with glues and compression techniques for shaping it to the desired form needed to manufacture comfortable chairs.

Their innovative architectural designs, which they both collaborated on, would look modern even in today's designs for open-plan living. But perhaps what is most surprising is that Charles and Ray delved into photography and film. They made short films to convey their ideas. Their films were works of art, and were intense studies of an idea or process.

It's hard to imagine how innovative the Eames must have been to create films like *Powers of Ten*, a video they created in 1977. This short film (available today on YouTube) shows a couple sitting on a blanket in a park. The blanket is 1 meter by 1 meter. The camera then zooms out by a power of 10 and shows the same couple on the blanket from 10 feet above the blanket, then zooms out again to 100 feet, and so on. Every 10 seconds it zooms out another factor of 10. It appears to go all the way out to the edge of the universe, then zooms in at the same pace to the couple, but continues on until it is close enough to show cells on their skin. This short film is remarkable to behold, and surprisingly avant-garde.

Charles and Ray Eames continued working together until Charles's death in 1978. After, Ray compiled a book titled, *Eames Design: The Work of the Office of Charles and Ray Eames.* She also archived all of their photography and sent it to the Library of Congress, which is where it still resides. Ray also continued projects as the head of the Eames Office, including the Eames Teak Sofa, hosted student group visits to the Eames House, and conitnued to give lectures about their work.

She died exactly ten years to the day after Charles did.

★

Very small teams applied creativity in other areas too, for example, photography and art.

Chapter 15
Creative Arts

Photography

A t the same time that Mackintosh and Macdonald were becoming well known in the design and architecture world, another pair of visionaries was operating in a completely different environment 10,000 miles (16,000 km) away. This team was working on expedition photography.

Scott and Ponting: Expedition Photography

In 1911, Captain Robert Falcon Scott was leading the multi-year British scientific *Terra Nova* Expedition to Antarctica, with the aim of being the first expedition to reach the South Pole, a trek of 800 miles (1,287 km) from the coast. A portion of this same expedition was described in chapters 11 through 13 about Wilson, Bowers and Cherry-Garrard's winter journey and their quest for emperor penguin eggs.

Scott sought out the best men with a wide variety of scientific expertise for his expedition, including the eminent photographer Herbert Ponting. Ponting was charged with creating a photographic record of the expedition. It was the first time any leader had brought the disciplines of science, discovery, exploration, and professional photography together into one coherent package.

In those days photography was a complex process of balancing focus, lighting and composition. The photographer had to understand these processes and bring them together to produce a single moment with the click of the shutter. Ponting was then, and still is now, considered one of the very best "camera artists."

Captain Scott was, at best, an amateur with a fair eye for composition, but had none of the other capacities. For him, having a visual record of his expedition was crucial to capture the scientific and geographical results, and of deep importance to the lectures and book publications that would help to pay off the expedition's debts. Ponting was to be his teacher in all these matters, and Scott was an ideal student.

The South Polar assault team consisted of Scott and four companions: Dr. Edward Wilson and Birdie Bowers (who we met earlier in this book), along with Captain Lawrence Oates and Edgar Evans. When Ponting returned home after one year in Antarctica, Scott became the stand-in photographer on the long trip to the South Pole and back.

Building on Ponting's tutelage, Scott took stunning photographs along the way. They reached the South Pole as planned in January 1912, but were not the first to the Pole; the Norwegian explorer

At the South Pole. *(from left to right)* Wilson, Bowers, Evans, Scott, Oates. *(Photographer: self-taken portrait.) (Library of Congress)*

Roald Amundsen and his team arrived five weeks earlier.[11]

Scott posed with his team at Amundsen's tent, left behind by the Norwegian to mark the location of the Pole. He took this iconic photograph — one of the first "selfics"– using a piece of string to release the shutter. You can almost see the despair and sadness on their faces at having arrived second, and in such a remote and unforgiving location. And they had an 800-mile-trek (1,287 km) back to the coast.

On the return journey, Scott's South Polar team met with disaster. Injury, frostbite, exceedingly cold conditions, and limited supplies in previously laid depots contributed to the tragic story. Scott and his men perished on the way back from the Pole.

Evans died first. A number of days later, Oates followed. Fearing his slow travel, due to his severely frostbitten feet, was endangering all of their lives, Oates left the safety of their tent, saying, "I am just going outside and may be some time." He never returned. Oates' sacrifice didn't save his companions. They all died on the return journey.

Their camera was recovered the following Antarctic summer along with Scott, Wilson and Bowers' bodies. They were found by a team of Scott's men including Apsley Cherry-Garrard. Scott, Wilson and Bowers were still frozen in their tent, pitched 11 miles (18 km) short of the next depot. Had they reached it, perhaps they could have survived the return journey. The bodies of Evans and Oates were never recovered.

In a surprising discovery many years later, a previously unknown cache of photographs showing polar scenery were sold at auction. This unexpected find turned out to be a collection of Scott's photographs taken during the tragic South Polar journey, under Ponting's tutelage. Scott the amateur had been well taught by the master Ponting. These newly discovered images were commemorated in David Wilson's 2011 book, *The Lost Photographs of Captain Scott.*

Because of Scott and Ponting's use of photography, further

---

[11] The story of Amundsen and Scott is told in our first book, *When Your Life Depends on It.*

expeditions of discovery were expected to include the very best photographic record of the new vistas. An example of this was several years later, when Ernest Shackleton hired the highly accomplished photographer Frank Hurley for Shackleton's famous *Endurance* Expedition. Hurley's work was already internationally renowned at that point. The best of his glass-plate negatives survived the sinking of the *Endurance*, and helped to tell the dramatic story of the expedition and the grueling rescue. (The expedition featured another small team: Shackleton and Frank Wild, his second-in-command. The Shackleton-Wild partnership is featured in chapters 37 to 40.)

Ponting's work with Scott was the precursor to the modern wildlife and outdoor photography documentaries created by luminaries like David Attenborough, which are enjoyed by millions.

Stieglitz and O'Keeffe: Photography as an Art Form

Ponting was an early proponent of what he called "camera art." Another pair of artists was making a name for themselves during the 1920s, creating visual legacies that helped define our sense of the twentieth century.

Georgia O'Keeffe at the time was a burgeoning artist. A friend brought some of her drawings to the attention of the photographer and well-respected New York City art dealer, Alfred Stieglitz. At his gallery, he innovatively displayed photographs in frames, and in a way that transformed photography into an art form that was in demand by collectors and museums.

Stieglitz and O'Keeffe met in 1916, fell in love, and married in 1924. Georgia was 29 when she met the older Alfred, who was nearly twice her age. Although together, they greatly influenced the art world, their long and fractured marriage was a marked difference from the happy union of Mackintosh and Macdonald.

Nonetheless, Stieglitz and O'Keeffe profoundly influenced each other and the art world of their era. Stieglitz is best known for his dramatic black-and-white images of New York City with their sharp lines and geometric juxtapositions of light and shadow. But he also photographed the natural world and the people inhabiting it. He was at his most productive after meeting O'Keeffe, creating over 300 photographs of her, some of them quite intimate. She was his model, and in a sense, his muse.

While earning her way through art school as a commercial illustrator, O'Keeffe developed her own unique painting style, starting with watercolor and charcoal images of natural objects rendered in an ever-increasing abstraction. She and Stieglitz often lived apart – he in New York, she in the arid lands of New Mexico. There O'Keeffe created large-scale, brilliantly colored – sometimes lurid – oil paintings of flowers and desert landscapes.

At their first show together in 1924, Stieglitz placed O'Keeffe's paintings alongside his photographs at the Anderson Gallery in New York. After, she began a series of dramatic urban abstractions with the city as her subject. He, in turn, found inspiration in nature.

★

The more we delved into the creative world, the greater variety of pairings we found. Each team member in these pairings was advancing not just their own art form, but the other's person's as well. Neither individual would have been as prolific or famous without the other.

The resulting creativity of these two-person teams inspired artists across their contemporary era, as well as all those who came after them.

This becomes even more evident in the areas of modern art and literature, as described next.

Chapter 16
Creative Arts

Modern Art and Literature

In the previous chapters we barely scratched the surface of some amazing couples like Mackintosh and Macdonald, the Eames, and Stieglitz and O'Keeffe, power pairings in the world of architecture, art and design. Other examples include the famous sculptor Rodin and his relationship to Camille Claudel; the Mexican painters Diego Rivera and Frida Kahlo; and the artist Christo and his wife Jeanne-Claude, best known for wrapping large buildings in cloth and other materials.

Christo and Jeanne-Claude: Modern Art

Christo and Jeanne-Claude were born on the same day in different countries. They met in 1958 and went on to create magnificent, large-scale art installations that were unforgettable by all who saw them. The scale of Christo and Jeanne-Claude's endeavors was magnificent. Each project took years, sometime decades to get planning permission, source materials and hire construction crews to erect them.

Their monumental work in 1976 involved erecting a 24.5-mile-long (39.4 km) "Running Fence" in California. It was made of white-nylon fabric hung from steel cables. In 1985, they wrapped the oldest bridge across the Seine in Paris in gold fabric; in 1995, they wrapped the Reichstag in silver fabric pinned to the building by blue ropes; and in 2005, they created 7,500 orange-cloth gates in Central Park, in New York City.

The scale of these projects was so enormous, it had to be experienced in person to understand and feel the splendor of the work. All of their projects were open to the public. They didn't have a ticketing system or set opening hours, so were free to be enjoyed like any statue in a public square.

Christo and Jeanne-Claude's pairing was a harmonious one, in contrast to Stieglitz and O'Keeffe. Jeanne-Claude died in 2009, but that didn't stop Christo from continuing their joint work. This included the epic project in 2016: the construction of huge, floating, yellow-cloth piers on an Italian lake. People were then invited to 'walk on the water'. It was a project that Christo and Jeanne-Claude had first conceptualized in 1970.

To give an idea of the scale of their projects, here are a few statistics about the floating piers. It required 1,000,000 square feet (93,000 square meters) of fabric sitting on 220,000 polyethylene cubes that floated on the water. Each cube was 52 feet (16 m) wide, and the

Some of Christo and Jeanne-Claude's 7,500 orange-cloth gates in New York City's Central Park. (Photographer: Brad Borkan)

walkway was almost 2 miles (3.2 km) long.

After Christo passed away in 2020, to commemorate Christo and Jeanne-Claude's joint work, the French government gave permission for the wrapping of the Arc de Triomphe in Paris in silver fabric, a project they had conceived in the early years of their relationship in the 1960s. The Arc de Triomphe remained wrapped for 16 days. It was a tribute to one of the greatest artist couples the world had ever seen. A very small team who had pushed the concept of art far beyond the scale of a canvas or a block of marble.

The next very small, creative team also worked in Paris and internationally.

Collins and Lapierre: Literature

Our inspiration to co-write this book came from the writing duo of Larry Collins and Dominique Lapierre. Their names may not be familiar, but their non-fiction books might be.

Collins was an American, educated at Yale and drafted into the US military in the 1950s. At the same time Lapierre was in the French military. They met a few times as a result of their military jobs, and later each became friends and journalists. Collins worked at *Newsweek* and Lapierre at *Paris Match*. Despite at times competing for scoops, they agreed to co-write a story of the French occupation. The result was a seminal work, titled: *Is Paris Burning?*

It tells the true story of how Hitler urged his men to blow up all the bridges in Paris as the Nazis were being pushed out of the city by the Allied forces, in September 1944. The Germans had loaded the bridges with explosives, but in the final days of the occupation, the head of the Nazi regiment in the city could not bring himself to carry out Hitler's order, despite the Führer constantly phoning him, asking, "Is Paris burning?"

It is a spellbinding tale of intrigue and sabotage told from American, French and German perspectives. It was a collaboration that could only have come about because of the authors' multinational partnership. The book, an instant bestseller, was turned into a major Hollywood motion picture.

After combining their investigative journalism skills, conducting detailed research, interviewing people from all sides, and applying a non-judgemental assessment of all the facts, they took on other weighty true stories, including: *Freedom at Midnight* about the separation of India and Pakistan in 1947, and *O' Jerusalem* about the formation of the state of Israel in 1948 and the Arab-Israel war that ensued.

It is likely that only a partnership of their professional intensity could have tackled these weighty and complex stories.

Legacy

All the small teams featured in these Creative Arts chapters – and surely there are many more – are excellent examples proving how pairings of bold, innovative artists working towards a common goal can push the boundaries of art, design and architecture. And this was achieved whether the individuals lived together harmoniously or not.

Our world has been immeasurably enriched by each of their pairings.

★

There are other creative, very small teams to consider, whose work in theatre, music and song continues to enrich our world. The most influential of these pairings was Gilbert and Sullivan. In their day, they were probably more popular and more influential than Elvis was in the 50s, the Beatles were in the 60s and possibly more than Taylor Swift is today.

Poster advertising Gilbert & Sullivan operettas Sorcerer, HMS Pinafore, and
Trial by Jury, performed by The Saville English Opera Company, 1879.
(Library of Congress)

Gilbert and Sullivan

"A Gilbert is of no use without a Sullivan"

—*W.S. Gilbert*

We started the previous set of chapters on the Creative Arts with the thought that the creation of art must be a solitary endeavor. We were fascinated to discover pairs of people pushing the boundaries of a variety of artistic endeavors.

We start this set of chapters on musical theatre with a completely different frame of mind. Musical theatre lends itself perfectly to pairs of collaborators – the lyricist who writes the words to the music, and the composer who creates the melodies to bring those words to life. Together, as a team, they create masterpieces far greater and more memorable than either could create on their own.

Famous pairings immediately spring to mind – Gilbert and Sullivan, and Rodgers and Hammerstein being the best known. But while Victorian-era Britain became known as the birthplace of musical theatre, it actually didn't start there, and it didn't start with a pair of people.

France and Britain in the 1800s

Opera had been popular in various European cities for centuries. By the 1800s, audiences in France were enjoying a variation on the opera called opéra-comique, which were satirical comedies blending music, spoken words, and songs.

In the mid-1880s, Jacques Offenbach, who was half French-half German, became one of the best-known creators of operettas in the opéra-comique genre. He wrote and composed over one hundred of them. Their popularity increased because they touched on a French theatrical interest at the time: bawdy plots, cross-dressing characters, and songs with sexual overtones.

At this same time, Britain was going through a transformation, thanks to two father-and-son teams. But that transformation was not in the theatre, it was in railways, and involved the Stephensons and the Brunels. Their work, however, had a direct impact on British theatre.

In 1829, George Stephenson and his son Robert invented the first locomotive, designed specifically to pull a passenger train. It was called the *Rocket*. Despite its name, the *Rocket* only averaged speeds of 12 miles per hour (19 km/h).

Marc Brunel was busy building the Thames Tunnel in London, the first tunnel ever built under a flowing river. His son, Isambard Kingdom Brunel, was the lead engineer working in the tunnel until he was seriously injured. Isambard went on to become one of the greatest engineers the world has ever seen, and his focus, too, became railways.

So, what does all this have to do with musical theatre?

Thanks to the sons, Robert Stephenson and Isambard Kingdom Brunel, competitors and collaborators to one another, railways proliferated across Britain. The Thames Tunnel enabled easier movement of people from one bank of the Thames to the other, but it was not until the Thames Tunnel was opened to rail travel in the 1860s that it became easier for Londoners to live farther from the city. Prior to the railways and the tunnel, people had to live close to where they worked. Now, they could commute by train from the rapidly growing London suburbs.

With this advent, families could travel into town for an evening's entertainment. The problem was London theatres had evolved from the style of Shakespeare's time. Modern theatre had become raunchier and debased entertainment to the level where the male actors were taken for ruffians and women performers for prostitutes.

With families seeking tamer and more family-friendly entertainment, a gap opened in the marketplace. Into the gap stepped three people who, as a team, not only transformed musical theatre, but had a profound and lasting effect on all theatre and entertainment throughout the English-speaking world, even to this day.

Richard D'Oyly Carte

Richard D'Oyly Carte spotted this gap. He was the enterprising manager of the Royalty Theatre in London at the time. He also had broad music skills and had set up his own agency as a talent scout, agent, and show producer. Carte was enjoying some moderate success bringing versions of Jacques Offenbach's French operettas to his theatre.

His desire was to make the operetta popular in England, and had decided British family audiences would warm to a different style than Offenbach. Carte wanted shows that were longer and less bawdy, but still possessing the satire, wit, and songs and music that Offenbach offered.

It was Carte whose vision and foresight brought the immense talents of Gilbert and Sullivan together to produce operettas. While now commonly referred to as Gilbert and Sullivan, they were specifically the lyricist, William Schwenck (W.S.) Gilbert and the immensely talented composer, Arthur Sullivan. Their first collaboration for Carte was *Trial by Jury*.

Trial by Jury

Carte recognized the superb comedy and wit in the text of Gilbert's newest play, and brought in Sullivan to compose the score. This appeared to be the start of a musical marriage made in heaven, at least from the theatre and operetta audience's point of view. The partnership, well that was another story, as will become clear.

Trial by Jury was an immediate and resounding success. In its first two years, it was performed over 300 times. The production kept going during those early years, even when it had to move into another theatre twice to maintain its long run.

The operetta's appeal was that it was filled with comic wit and memorable tunes about a lawsuit being tried in a courtroom. The premise is Edwin is on trial for having broken his promise to marry Angelina. The cast of characters includes the judge, jury, Edwin (who has fallen in love with another woman) and his jilted bride-to-be Angelina, as well as Angelina's bridesmaids. Angelina's beauty charms the jury and judge.

What Carte recognized better than most was that mockery of important people and institutions like the judge and the court of law appealed to British audiences. Hidden within this tale is an underlying satirical takedown of societal norms and people in high positions who might appear to be respectable but are not. *Trial by Jury* ends with the judge being frustrated by the jury's inability to decide, and is so charmed by Angelina that he decides if Edwin doesn't want to marry her, he will. Angelina accepts.

What Gilbert and Sullivan had expertly displayed in this theatre show (which ran in only one act, lasted less than one hour and could be staged within an evening of other shows), was how the audience loved seeing the hypocrisy and duplicity of important people woven into the fabric of the story on stage. Audience members who might have felt powerless against the ruling class of British society could come away from an evening's entertainment content in the knowledge that their leaders deserved, and were on the receiving end, of a certain amount of ridicule.

Opening night reviews described *Trial by Jury* as fun, humorous, amusing and original. The staging added to the hilarity. There was something in it for everybody, whether you loved the storyline, the music, the subtle and not-so-subtle jokes, as well as the innuendos and double entendres. The happy, light-hearted ending, with the judge proposing to Angelina, ensured the audience left entertained and wanting more. Gilbert and Sullivan came out to rapturous applause on opening night.

Gilbert and Sullivan's start

Gilbert and Sullivan knew each other before D'Oyly Carte paired them up on *Trial by Jury*. Several years before they had collaborated on a Christmas show. Though both were British, their backgrounds and temperaments differed.

W.S. Gilbert was born in 1836, and there is a story that when he was two years old, he was kidnapped while on a family vacation to Italy. The ransom of £25 was paid by his parents. He was educated at Oxford and trained as a lawyer. He had a keen interest in comedic writing.

Arthur Sullivan was six years younger than Gilbert and at an early age was an exceptionally talented musician. At age 14, he was awarded the Mendelssohn Scholarship to attend the Royal Academy of Music. He came to prominence with his graduation piece. From there, he went on to compose many pieces including a symphony, as well as music for the church. Even if he hadn't collaborated with Gilbert, he'd still be renowned for the music of *Onward Christian Soldiers*.

Gilbert was 39 years old and Sullivan was 33 when Carte paired them up for his purposes. Arthur Sullivan was a genius (he was sometimes called England's Mozart), and W.S. Gilbert could have fitted

W.S. Gilbert, c1882 *(Top)* and Arthur Sullivan, 1867.

into the genius class as well, but their upbringing and education meant they wore their great talents differently. It seemed that, as a team, they were always destined to come apart. The only question was when and what would cause it.

Following the success Carte had with *Trial by Jury,* he launched the Comedy Opera Company. His original intention was to work with a roster of lyricists and composers but dropped that idea and focused on teaming with Gilbert and Sullivan.

Did they like each other?

Gilbert and Sullivan didn't really like each other, but they both acknowledged the power of their combined partnership. One can imagine the power struggles that ensued, where the lyricist wants the composer to adjust their music to the words provided, and the composer demands that the words be adjusted to fit the tempo, rhythm and structure of the music.

Despite strong differences, they were able to spur each other on to reach their individual potential. This may be one of the most salient lessons from all the teams we studied for this book – that liking one another is not essential for a team to reach greatness.

Together they produced 14 operettas across 25 years of collaboration. The most famous of these were: *HMS Pinafore* (1878), *Pirates of Penzance* (1879), *Iolanthe* (1882), *The Mikado* (1885) and *The Gondoliers* (1889).

Gilbert and Sullivan started blurring the lines of what was an operetta, moving towards musical theatre. Each of their shows had similar core elements that blended satire and wit with spoken words, and songs infused with catchy melodies. Some of their shows were so popular they were appearing at a London West End theatre at the same time as being staged in local or regional theatres throughout Britain. They also started being performed in the United States.

At The Savoy Theatre

In the late 1870's, while Gilbert and Sullivan were clashing and still producing their best work, Carte was building his own theatre specifically for staging their operettas. The Savoy Theatre, located on a busy London Street known as the Strand, opened in 1881.

The first performance there was a Gilbert and Sullivan creation called *Patience,* which is about two suitors – one who pretends to be a poet and the other a real poet – vying for a young woman's affection. Like all their plots it is far more complicated, and involves women of the town and a regiment of military officers.

But what is far more significant than the play is the theatre itself. It was considered the most beautiful theatre in Europe, with numerous innovations introduced by Carte and his theatre manager. For example, it was the first theatre to sell tickets for assigned seats. The attendants were paid fair wages so no tipping was required for the cloakroom, and programs were distributed for free.

Even more importantly, it was the first public building in the world to be lit entirely by electricity. (The electric light bulb had only been patented two years before.) All other theatres at the time were lit by gas, which was not only a fire hazard, but it created an odor and increased the heat in the building.

The next year, four weeks before Christmas in 1882, *Iolanthe* was staged at the Savoy Theatre. Like all of Gilbert and Sullivan's operettas, to summarize the plot in just a few sentences does not do justice to the nuances, humor, fun and satire of the show. The gist of *Iolanthe* has to do with a fairy named Iolanthe who married a mortal, and now has a son named Strephon. The young man's upper half is fairy, and his legs are mortal. Strephon wants to marry a young mortal woman named Phyllis (a deed carrying a punishment of death). But Phyllis is not an ordinary mortal. She lives with her guardian, the Lord Chancellor, head of the British Parliament's House of Lords. Somehow things get resolved between the fairy world and the House of Lords.

Because the Savoy was fully electric, *Iolanthe* could be more cleverly staged and lit differently than any previous theatre production ever produced. For example, Iolanthe and the principal fairies wore battery-powered star lights in their hair. The term 'fairy lights' actually came from the staging and costume design of *Iolanthe*.

The show opened to great acclaim and was reported to have had nine encores. *Iolanthe* emerged as another Gilbert and Sullivan hit, thanks also to the entrepreneurism of D'Oyly Carte. The Savoy Theatre became the launch pad for every future Gilbert and Sullivan operetta created, resulting in their works and those of imitators being called Savoy Operas.

Gilbert and Sullivan's success was so closely tied to Carte's own brilliance that it is somewhat reminiscent of the Wright brothers' success and links to Charles Taylor, who built their lightweight aluminum airplane engine.

More successes, then...

Gilbert and Sullivan went on to produce more operettas, which were known as comic operas. They poked fun at the establishment and politics using wit, humor, comic timing, costumes, scenes and satire. They did this often in the shows that came before *Iolanthe*. In *Pirates of Penzance* a key character is the Major General, who was supposed to be a military leader but knew almost nothing about warfare or technology. Their show *HMS Pinafore* was a love story that parodied the British upper class, showing how unqualified people from that sector of society rose inappropriately to positions of authority.

But after *Iolanthe*, which some might have considered to be the peak of their prowess as a team, they created *The Mikado*.

The Mikado was set in Japan. Nanki-Poo was a young man in love with Yum-Yum, but she was already engaged to Ko-Ko, the town's tailor. Ko-Ko had been accused of flirting, an offense punishable by beheading, but the town officials, who were against such barbaric punishments, decided to give Ko-Ko a reprieve by giving him the executioner's job. In other words, to fulfill his job, he must behead himself first.

The Mikado, who was emperor of Japan and Nanki-Poo's father, was a leader who actually did want the beheadings to occur. Ko-Ko was unsure of what to do until he discovered that Nanki-Poo was so distraught over losing Yum-Yum, he was considering suicide. Ko-Ko suggested that Nanki-Poo take his place and be executed later in the month. Nanki-Poo agreed on the one condition that he got to marry Yum-Yum for one month. This plan failed when they realized the Mikado had previously declared that the wife of any beheaded man would be buried alive.

Ko-Ko agreed to fake Nanki-Poo's execution and when the Mikado arrived, he was told about Ko-Ko's first and only 'execution' on the job. The Mikado was also looking for his son, Nanki-Poo, but to produce him, Ko-Ko needed to admit he lied to the emperor.

As with all Gilbert and Sullivan's productions, the plot is way more involved and has more characters than can be summarized in a couple of paragraphs. Among the many appeals of *The Mikado* are the delightful character names Gilbert and Sullivan breathed life into. A character named Pooh Bah was designated as 'the Lord High Everything Else'. He also possessed other high-sounding titles above his level of ability. Pooh Bah's self-confidence, mixed with his incompetence, resulted in his name being immortalized with the expression, the Grand Poobah, a comic and derisory title given to anyone with those characteristics.

The Mikado was a hit from opening night onwards. According to an article by the cultural critic H.L. Mencken in the *Baltimore Evening Sun*, the show opened on March 14, 1885 and by December 31 of that year it was being staged by 150 theatrical companies in Europe and the US. Mencken claimed that it was so popular on a specific October evening in the US there were over 117 performances of *The Mikado*. Without copyright protection in the US because the show was written in Britain, Gilbert and Sullivan and Carte were not paid royalties for foreign productions.

For those of us who can remember Beatlemania, when the Beatles, led by another songwriting duo Paul McCartney and John Lennon, were in their heyday, there was a frenzy of excitement about everything they did, most especially their first visit to the United States. Gilbert and Sullivan may not have had such a fanatical personal following, but *The Mikado* certainly had that level of fame.

Mencken wrote in his article that people in the US were "*Mikado* crazy for a year or more." Americans became fascinated with the people and customs of Japan, their style of dress, food and other details.

The breakup

After *The Mikado*, came two more operettas and then another big hit in 1889, *The Gondoliers*. This coincided with D'Oyly Carte opening the Savoy Hotel in London next to the Savoy Theatre. The hotel was the first luxury hotel in the world, and featured Art Nouveau décor and the first bar to serve cocktails in the city. It was frequented by celebrities, royalty, and elite people from around the world.

It was reported that in 1880, which is before their hit operettas like *Patience*, *Iolanthe*, *The Mikado* and *The Gondoliers* were even written, Carte, Gilbert and Sullivan were generating £60,000 in annual profit (approximately $10,000,000 today). What they were earning at the time of their breakup must have been considerably more than that, and that is what makes their breakup so perplexing.

Sullivan, the composer, always took a keen interest in the finances of the Carte-Gilbert-Sullivan endeavors. He was taken aback by the expenses required for *The Gondoliers* which was £4,500 and when he queried Carte on this, he discovered £500 of that money had been spent on new carpets for the lobby of the Savoy Theatre, which was Carte's and not owned by Gilbert and Sullivan.

Sullivan had already been thinking about moves to create grand opera not comic operettas, and Carte liked the idea. However, a reconciliation over the carpet cost could not be overcome, and after a messy court hearing involving a judge (the irony is that their first big success, *Trial by Jury*, focused on a legal case and a judge), the three decided to split from one another.

A few years later Gilbert and Sullivan co-wrote two more operettas, but neither was as successful as their earlier work.

Their legacy

Gilbert and Sullivan had an immense and profound impact our culture. Their influence on musical theatre extends all the way to the modern show *Hamilton*. In the first act of that show, George Washington exclaims that he is, "The model of a modern Major General." At that moment in *Hamilton,* Washington's Revolutionary War army is out-gunned, out-manned, out-numbered, and out-planned, and Washington's character is implying he feels as incompetent as the buffoonish character called Major General in *Pirates of Penzance.*

In addition to influencing Lin-Manuel Mirada, the author of *Hamilton*, Gilbert and Sullivan inspired other great playwrights including Noel Coward, Andrew Lloyd Webber, Stephen Sondheim, and many others.

Phrases from Gilbert and Sullivan shows are widely known and often used, such as 'Let the punishment fit the crime' from *The Mikado*, and 'A policeman's lot is not a happy one' from *Pirates of Penzance.*

Their work not only spun off books, films, TV shows, songs and other entertainment directly based on the plots, lyrics and music, but it has influenced many other areas of endeavor, where wit and satire have been used to lampoon leading figures.

Television shows and films have borrowed themes, songs, and sentence and lyric-fragments from Gilbert and Sullivan productions. *Raiders of the Lost Ark* is a good example. As Indiana Jones and Marion Ravenwood get ready to escape the Nazis by boarding a ship that Jones' friend Sala has arranged for them, Marion gives Sala a goodbye-thank you kiss. As they leave, an elated Sala bursts into a song, "A British tar is a soaring soul, as free as a mountain bird."

Tar is slang for sailor. The song is from *HMS Pinafore.*

Thanks to Gilbert and Sullivan's broad appeal to people of all ages and backgrounds, their work has been adopted by amateur and semi-professional dramatic societies around the world. It has helped expand local and regional theatres because their operettas can be easily adapted to smaller venues. Acting and production teams with limited budgets for stages, sets, props, and costumes can stage the plays, letting the satirical word-play and well-known songs carry the evening.

A duo or a trio

The genius of Gilbert and Sullivan is fully on display in every show. They are rightly applauded as the duo that transformed musical theatre. But D'Oyly Carte's role in bringing them together, and keeping this disparate pair working – if not in harmony, at least toward a common goal and greater good – cannot be underestimated.

Just like the important role Charles Taylor played with the Wright brothers, and Barry Sussman played with Woodward and Bernstein (discussed later in Chapter 20), Carte's endeavors cannot be underestimated. Gilbert and Sullivan may never have come together, created the works they did, or enjoyed the same level of success had it not been for the third person on their team, Richard D'Oyly Carte.

D'Oyly Carte's Savoy Theatre built in 1881, where every Gilbert and Sullivan show opened, is still standing and operating as a theatre, though it has been through multiple renovations since. It still hosts musical theatre productions, all of which are likely to have been influenced in one way or another by the talent, wit and satire exhibited in the works of the team Gilbert and Sullivan.

Where their work had the most influence was on the pairs or small teams in musical theatre that came after them. These include greats like Rodgers and Hammerstein, and George and Ira Gershwin. Their collaborations are showcased in the next chapter.

Rodgers, Hammerstein and Others

The most famous musical theatre pairing of a lyricist and a composer after Gilbert and Sullivan was Rodgers and Hammerstein, but theirs was not a strictly linear progression. Richard Rodgers (composer), and Oscar Hammerstein II (lyricist), each had been part of a previous pairing that had led to truly notable musical successes.

Hammerstein and Kern: *Show Boat*

Oscar Hammerstein II started partnering with Jerome Kern in 1925. Hammerstein was 30 years old. Kern, ten years older, was a New York-based pianist and song writer who had trained in the US and Europe. Their first collaboration, *Sunny,* did very well.

Their next team effort would change musical theater forever.

Kern had a goal: to turn a wide sweeping novel that followed the lives, loves and challenges of a three-generation family from the 1880s to the 1920s into a Broadway musical. The book, Edna Ferber's best-seller *Show Boat,* followed this family up and down the Mississippi River on their own itinerant floating theater, the stern-wheel paddleboat called *Cotton Blossom.* The book dealt with complex topics not previously covered in operas, operettas or other musical variety shows of the time. The topics included issues like gambling and alcoholism, racism, debt repayment, interracial marriage, and family breakdowns.

Up to this point, musical theater had come from the satire, wit and comedy of Gilbert and Sullivan. While we can marvel at the originality that Gilbert and Sullivan had achieved, their work was largely fanciful. *Show Boat* was on a completely different level. Kern and Hammerstein's collaboration as a team resulted in the most transformative stage show ever written.

It had both serious and comedic elements, with the story told through song, verse and spoken word. Rather than telling the story in the lighthearted manner of a Gilbert and Sullivan operetta, Hammerstein and Kern gave real human depth to the characters, bringing them to life in a way that allowed the audience to feel real empathy and compassion for their plight and their future.

This was a saga set to song, and it brought audiences through a whole range of emotions from funny and sad, to poignant. According to those who attended the first shows, unlike other theatre events where applause was immediate after the final curtain fell, at the end of *Show Boat*, audiences sat in stunned silence for a couple of minutes before applauding.

They had literally never seen anything like it.

Unlike Gilbert and Sullivan productions with their distinctly British orientation, *Show Boat* was uniquely and fully American – a rolling, epic saga with American angst, challenges and the quest for the American dream. With themes like racism, the cast was multi-racial – a first for any big-name show at the time.

Jerome Kern's score was about more than just songs; his music brought the scenes to life. Songs from *Show Boat* became standalone hits, like *Ol' Man River,* and *Can't Help Lovin' Dat Man,* but the whole show worked because it was thought of as one entity – one show with a story arc spanning years and across many locations.

It is hard to emphasize how innovative this production was in 1927. Miles Krueger, the president and founder of the Institute of the American Musical, said that there were two distinct eras to the American Musical: everything before *Show Boat* and everything after.

Hammerstein and Kern continued to collaborate as a team

creating a variety of musical theatre shows until Jerome Kern died from a cerebral hemorrhage in 1945, with family members and Oscar Hammerstein II by his side. *Show Boat* was twice made into a Hollywood movie using the songs of Hammerstein and Kern; once during Kern's lifetime, and once afterwards.

Rodgers and Hart

Even before Hammerstein and Kern partnered together, Richard Rodgers was teaming up with Lorenz Hart in 1919, when Rodgers was just 17 years old. Hart was several years older. They decided to work together as soon as they met. Rodgers later said of that fortuitous first meeting, "I left Hart's house having acquired in one afternoon a career, a partner, a best friend, and a source of permanent irritation."

They worked together for 24 years writing over twenty musical comedies, despite having very different temperaments. Rodgers was organized, controlling and met deadlines. Hart was the opposite, but they both strived for perfection and would argue often over lyrics and composition.

Like all theatre writers and composers of that era, their own work was influenced by *Show Boat*.

Their most memorable songs, the hits *Blue Moon, My Funny Valentine*, and *Bewitched, Bothered and Bewildered* remain favorites to this day. On Broadway, their most famous shows included *The Boys from Syracuse* in 1938 (probably the first time a Shakespeare play had ever been turned into a musical comedy), and *Pal Joey* (about a male nightclub performer with desires of grandeur who seduces a wealthy widow), which opened in 1940 with Gene Kelly in the lead role.

Hart's use of alcohol affected their ability to work together, and by the early 1940s, Rodgers was seeking an alternative lyricist. He had a project in mind called *Oklahoma!,* but felt Hart was too erratic to work on it. By 1943, Lorenz Hart's drinking was out of control, and he died later that year of pneumonia.

Fourteen years later, Rodgers and Hart's *Pal Joey* was turned into a film starring Frank Sinatra and Rita Hayworth. The Rodgers and

Hart song *Blue Moon* was so popular that Elvis Presley, and then later Bob Dylan, each recorded a version of it.

Rodgers and Hammerstein

With the need to find a lyricist for *Oklahoma!*, Richard Rodgers had Lorenz Hart's blessing to turn to Oscar Hammerstein II.

Oklahoma! was another breakthrough, transformative production of musical theatre, not as monumental as Hammerstein and Kern's *Show Boat*, but nearly. It was set in the Indian and Oklahoma territories in 1906, one year before President Theodore Roosevelt joined the regions together to form a US state.

Like *Show Boat*, it is a deeply American story based on a book. The premise is a love story between a farm girl named Laurey, and Curly a desirable, attractive cowboy. A brutish ruffian named Jud also seeks to win Laurey's affections. Meanwhile another love triangle is playing out with Laurey's boy-crazy female friend Ado Annie, and her two male suitors. Various threats and fights result, and all ends well at least for most of the people involved.

It was not the plot that was the breakthrough, but the way Rodgers and Hammerstein crafted the show to be an interwoven tale between likeable and unlikeable characters told through songs, dialogue, costumes, sets and stage props, long dance numbers, comedy, fight scenes, and love stories. Themes and elements ran throughout the show giving it continuity and dramatic flair, from the opening number to the finale. Even more than *Show Boat* and unlike anything that had gone before, it was drama set to music and dance.

With the songs center stage, the casting for the show focused on finding singers who could act, rather than actors who could sing. This gave tremendous power and enthusiasm to the opening number *Oh What A Beautiful Morning*, a well-known hit in its own right.

Rodgers and Hammerstein hired choreographer Agnes de Mille to design the dance sequences. Her work was revolutionary in that the dances she created for *Oklahoma!* were not only integral to but advanced the storyline. They helped to show what characters like Laurey, Curly and Jud were thinking and feeling.

The show opened in 1943, when the US was immersed in World War II. A show celebrating rugged Americanism, love, and the triumph of good over bad was the perfect anecdote for the country.

Oklahoma! ran for over five years on Broadway, breaking all previous records for a show. It ran for 2,200 performances (four times greater than any previous Broadway run) and won a special Pulitzer Prize. The Tony Award had not yet been created.

In 1953, the state of Oklahoma adopted the theme song from the show as its state anthem.

Working style

From the start, Rodgers and Hammerstein developed a collaborative style that worked best for them. Hammerstein would write the lyrics to a song, and then Rodgers would compose the music. In contrast, Rodgers and Hart had worked more closely together, but that also led to contentions because neither wanted to change what they had written or composed to suit the other.

Although Rodgers and Hammerstein both had homes in New York City, they spent considerable time in other states, Rodgers at his second home in Connecticut while Hammerstein lived part of the time in rural Pennsylvania. There he could work slowly and carefully on the lyrics and then send them to Rodgers, who set them to music. Since Rodgers already knew what the song was about, who would be singing it, and where it fit in the plan for the show, he could compose the score quite quickly.

The power of teamwork

Rodgers and Hammerstein went from one big-name success to another big-name success. These included *Carousel, South Pacific, The King and I,* and *The Sound of Music,* and many more. They lived productive lives, receiving many accolades that are testament to their ability to team together for the greater good. This included winning over 40 Tony Awards, 15 Academy Awards, two Grammys, two Emmys, and two Pulitzer Prizes.

Further proof of the enduring quality of the work is the number of revivals their shows continue to have. Even as we write this today, *Oklahoma!* is on stage in London, *The King and I* is on a world tour, and it is likely that some local theater in the world, be it student, amateur, semi-professional or professional, is staging one of their shows at this very moment.

Long after both members of one of the greatest musical theater teams the world has ever seen, passed away, *Time Magazine* placed them in their list of the top 20 most influential artists of the 20th century.

Neither could have achieved that on their own. It was all due to teamwork.

George and Ira Gershwin

We have only scraped the surface of these great teams and what they accomplished. Another great team in musical theatre is that of George and Ira Gershwin. They lived and worked in the 1920s and 1930s, at the same time as Rodgers and Hart, and Hammerstein and Kern.

The Gershwins were brothers. George was a musical genius who, at the age of 11, learned to play piano by following the depression of keys made by a player piano. George rose to fame thanks to the hit song *Swanee,* made popular by Al Jolson, and his lengthier compositions *Rhapsody in Blue* and *An American in Paris.*

When it came to creating musical theater, his older brother Ira became his lyricist. Together they crafted many musicals like *Crazy For You,* and *Porgy and Bess.* When it first opened, *Porgy and Bess* did not gain the fame and recognition it has today.

Unlike Rodgers and Hammerstein, where the lyrics were written first, due to George's musical talent, he would write the melodies first, and Ira had to fit his words to the tempo of the music. This is a more difficult task than the other way around. At the peak of their success, George succumbed to illness and died of a brain tumor when he was only 38 years old.

A richer, fuller world

The transition from opéra-comique to operettas to musical theater has made the world a richer, fuller place. And the beauty of it is, none of the earlier shows are lost.

These came about thanks to immensely talented and hardworking teams like Gilbert and Sullivan along with Richard D'Oyly Carte, Rodgers and Hart, Hammerstein and Kern – moving on to Rodgers and Hammerstein – and George and Ira Gershwin. Each left a distinctive mark on the genre.

An almost straight line can be drawn from the pathbreaking shows of Gilbert and Sullivan to Hammerstein and Kern's *Show Boat,* to Rodgers and Hammerstein's *Oklahoma!,* to today's most modern and successful transformative show, *Hamilton.*

One of the lessons we can learn from looking at these teams and their work is that team members don't have to like each other to create great outcomes as Gilbert and Sullivan did, but it helps if you do, as proven by Rodgers and Hammerstein. And it doesn't really matter if the lyrics or the music is written first or second, the key is the vision of the work and the focus on its completion.

★

In the next chapter, we turn to a different type of creative teaming – not one based on satire or comedy but on real-life events. It started with two young journalists, neither of whom had much time for the other, competing on a small crime story.

It ended with an outcome that shocked the world.

The United States White House. (*Photographer: Brad Borkan*)

Chapter 20
Investigative Reporting

Woodward and Bernstein

"I hereby resign the Office of President of the United States."

—*Richard Nixon, August 9, 1974*

Not all very small teams come together by choice. The Eames were married to each other; Orville and Wilbur were siblings; Peary and Henson started out as employer-employee. But two people, occasionally joined by a third – whether by chance or by prior relationship – may discover they share an interest in solving an age-old human problem.

Unlike the other teams, Bob Woodward and Carl Bernstein were not motivated by a common goal of their own choosing. Barely acquainted with each other, they were two of the lowest ranking, most newly hired reporters in the newsroom of an esteemed newspaper, the Washington Post.

In 1972, they were assigned to research a news story by their senior editor. Collaborating and working on an obscure story was not a goal either would have chosen. There were far more interesting and newsworthy events swirling around Washington, D.C. in June of that year than an insignificant, failed break-in at the Democratic Party's campaign headquarters on the fifth floor of the Watergate Hotel.

Two of the most important stories in the first six months of 1972 were President Nixon's eight-day visit to China, and his historic meeting with China's leader Chairman Mao. Its aim was to normalize relations between two of the most powerful nations on the planet. The second was the US involvement in the politically divisive Vietnam War and the shocking, just published photograph of a nine-year-old Vietnamese girl running down a road screaming, her clothing burned away during a napalm strike on her village.

In comparison to these, an attempted burglary in a luxury Washington, D.C., hotel and office complex hardly mattered.

Rivalry

Woodward and Bernstein, both young reporters in 1972, were assigned desks in the same row – separated by 25 feet (7.6 m) and a building column – in the vast newsroom of the *Washington Post*.

Although nearly the same age (28 and 29) and marital status (one was separated; the other divorced), they had entirely different upbringings. Woodward was the scion of a patrician Republican family, son of a judge, and raised in a world of country clubs and relative ease. He attended Yale University. Bernstein was the scrappy Democrat – a college dropout who left university due to poor grades, and became a self-made reporter accustomed to following his own instincts, He had been in the game since he was 16 years old.

Neither had much use for the other.

Despite the proximity of their desks and their ages, no friendship bloomed between them. They had never worked together before now and neither was particularly happy with the arrangement.

Woodward's first thought was, "Oh God, not Bernstein," recalling the other reporter's reputation around the newsroom for pushing his way into a good story. Bernstein had heard that Woodward's rapid rise at the *Post* was a result of his elite societal upbringing and status.

Given the choice, neither would have worked with the other or even found a single reason to develop a friendship.

A burglary

It came as a surprise to both of them when they were independently assigned to work on the same story about a burglary at the Democratic campaign headquarters, in the elegant and exclusive Watergate Hotel.

The presidential campaign season for the 1972 election was heating up. Republican Richard Nixon was planning to win his second term in November. George McGovern, his opponent, for all his progressive anti-Vietnam War credentials, was not seen as a strong enough competitor for the Committee to Re-elect the President to worry about much.

Since it was looking like Nixon would be a shoo-in for re-election, this lead about a botched burglary of the Democratic headquarters seemed like a minor story.

The Watergate complex with the Watergate Hotel on the right
(Photograph: Angela and Michael Galper)

The *Washington Post's* police beat reporter Alfred Lewis broke the story on June 18. The headline read *Five Held in Plot to Bug Democratic Offices Here.*

The senior staff at the *Post* believed there must be more to be reported, and assigned Woodward and Bernstein to each write separate stories, only one of which would likely end up in print. The editors did not tell the two reporters they were competing.

Was this a gambit to pit them against each other to stoke the natural competition between reporters for a byline?

When Woodward handed in the first three paragraphs of his story draft, within minutes he noticed Bernstein hovering over the city editor's shoulder making suggestions, then taking the draft back to his own desk to start rewriting it. Minutes later, Woodward handed in the second page, and minutes after that Bernstein was at his desk, typing again. Woodward walked over to find out just what was going on. Reading the revision, he had to admit – it was better.

Their story went into print the next day, with a more explosive headline than the first: *GOP Security Aide Among Those Arrested.* One of the five burglars was James McCord, and it was revealed that McCord was on the payroll of the Committee to Re-elect the President.

Though neither of the reporters could know it, this marked the beginning of their work as a team. Woodward and Bernstein were, after all, but two cogs in the furiously spinning machinery of one of the greatest metropolitan newspapers in the country. The *Washington Post*, served up more than just a recounting of each day's newsworthy events. It was the nation's principal watchdog over the activities of the United States government – the legislation produced in the Senate and House of Representatives, the cases before the Supreme and District courts, and the Presidency – but also the sometimes-shady underbelly of the electoral process in all its complexity. Everyone in government knew that the *Post* was watching even if no other newspaper was, digging deep in its relentless search for truth, and writing it up for all the nation to see.

These two young men had started near the bottom of a news staff of 378 people, handling whatever assignments they were given.

The teamwork that they developed over those early months on the Watergate case arose less from their own initiative than from commands from their editor. Both these men had already established themselves as solid professionals in the field. The *Post* would not have hired them otherwise.

But what did it take to make a good reporter in the 1970s? A nose for news, an overflowing rolodex of names and phone numbers, and a dogged search for the facts of a story, however much work and even at times risk, it took to get them.

Woodward was the faster writer and typed out most of the first drafts; Bernstein was the better writer, revising and polishing his partner's work late into the night before handing it in to the daily paper's editors. Thrust together during the steamy Washington, D.C. summer of 1972, the pair gradually overcame their mutual distrust and suspicion. They realized there could be real advantages to working together as a very small team.

For the Watergate burglary story, two reporters could be required. It had the potential to become too broad, with the stakes growing greater as their investigations could reach into the inner workings of the Nixon White House. There were heady risks involved, for them as well as the clandestine contacts each man had. They could divide the work and share resources.

Or there was another option: become friends.

Honing their craft

Rules had to be followed as Woodward and Bernstein pursued their leads into the tangled web of what appeared to be an insignificant break-in.

Post editors Barry Sussman, Harry Rosenfeld, Howard Simons, and executive editor Ben Bradlee were part of the greater team that ensured the highest journalistic standards were met. This was especially true for the growing reams of news stories that probed ever deeper into the workings of the US federal government.

Sussman, with his innate capacity to see the most relevant facts of a story and file them away in his memory, became a living encyclopedia of everything Watergate. He was a reference to be consulted when all other avenues failed. More than any other person in the newsroom, he understood how all the pieces of the puzzle fit. Sometimes he would join Woodward and Bernstein in the office late at night, after his own daily editorial duties were done, going over the details of their work, looking for clues and leads as to the next direction of their investigation.

Journalistic integrity mandated that no story could be run without corroboration from at least two independent sources, and for many sources, confidentiality was paramount. Every story Woodward and Bernstein wrote had to meet the exacting standards imposed by the *Post's* editors which required them to track down every potential lead and angle. If all the pieces didn't fit perfectly well together, the story would not be printed. Impressed by the work and determination he saw from the pair, executive editor Bradlee put the reputation and the resources of the *Post* to aid his very small team of investigative bloodhounds.

Digging deeper into the story, Woodward and Bernstein found evidence that, shockingly, seemed to lead upward toward the White House. Like any competitive reporter, each thought that the other would walk off with the byline for the story, and get to bask in the glory and recognition that would come with it. Rather than let that happen, they each worked even harder.

If one spent the weekend chasing down a lead, the other felt he had to do the same. They submitted a front-page headline story titled *Bug Suspect Got Campaign Funds* for the August 1, 1972 edition of the *Post* about some mishandled cash from a slush fund controlled by the White House. It claimed that one of the burglars, James McCord, had received money from Nixon's campaign. Both reporters were named in the byline.

Bernstein was in Miami when Woodward wrote the follow-up story for the next day's *Post*. He insisted that both their names appear under the headline.

The story grew even bigger on October 10 with the headline: *FBI Finds Nixon Aides Sabotaged Democrats*.

This news did not stop voters from handing Nixon a landslide victory in the general election on November 7. Nixon and his Vice-Presidential candidate Spiro Agnew were wildly popular. The Nixon-Agnew ticket won over 60 percent of the popular vote, securing 49 of the 50 states, and won over 96 percent of the Electoral College votes. It was a decisive command performance. In comparison, McGovern – the liberal who wanted to end the Vietnam War and institute social welfare reforms through taxing wealthy Americans – won only Massachusetts and the non-state of Washington, D.C.

The presidential election may have been a competition, but Woodward and Bernstein's early competitive style in working against one another was evolving. They were becoming a highly functional very small team.

Chapter 21
Investigative Reporting

Learning To Work As A Team

From that point on, all their reporting – and there was a lot of it – with far-reaching consequences was published under the shared byline. In the newsroom, their colleagues called them "Woodstein."

As their mutual distrust diminished, they came to see the advantages of working together. Each had different instincts and talents to bring to bear in a cooperative and complementary teamwork analogous to the working relationship of Susan B. Anthony and Elizabeth Cady Stanton in their quest for women's rights in the mid 1800s.

While chasing down leads in the field, Woodward's calm demeanor and familiarity with conservative values made him more appealing to the well-heeled supporters of the Nixon administration. Bernstein's man-of-the-people persona helped him get in with those who distrusted the president and the inner workings of the White House.

Each kept his own list of contacts, so important in a field where trust must be maintained. Sometimes Woodward and Bernstein met with and interviewed leads together, sometimes separately, but as their partnership blossomed, they always shared the results of their investigation.

Each would write their own first drafts of the Watergate-inspired news stories and trade them back and forth, rewriting their own and each other's words, until they felt they had the story right – well-told, scrupulously accurate, and highly protective of their sources of their information. Often, while working on a story well into the night, they would argue violently over the choice of a single word or sentence, to get the nuance of the story exactly right.

Their ability to vehemently argue points without damaging their relationship was reminiscent of the Wright brothers' intense arguments over glider and airplane design. Great teams it seems have the ability to do that and produce even more significant outcomes.

Evolving as a team

Woodward and Bernstein's book *All the President's Men* recounts their Watergate journalism in great detail, naming so many names and disturbing activities that the casual reader today may get overwhelmed by all the elements. But at the heart of their book lies the origin of one of journalism's strongest teams.

Woodward and Bernstein did not always function like a well-oiled machine. They worried about each other's social proclivities. Bernstein thought his partner didn't have the street smarts to stay out of trouble. Woodward thought Bernstein was all too casual in questionable company.

There were missteps along the way, some of which threatened to undo all the careful work they had already accomplished. The secretive nature of their enterprise made room for plenty of miscommunications between them. They had near-disasters resulting from over-optimistically weaving tenuous threads of information together.

At times, they arrived at spurious conclusions derived from circumspect conversations with overcautious contacts. This was not surprising. Sources and contacts were scared about the direction Woodward and Bernstein's growing investigation was taking and feared jeopardizing their own or their superiors' employment by revealing too much. One of the many challenges Woodward and Bernstein faced was discerning facts from obfuscation or lies.

The investigation deepens

One man, known only by the code name "Deep Throat" became one of the most trusted and useful sources, and one of the most protected. It was he who met with Woodward in a darkened parking garage; it was he who provided how best to proceed with the investigation, and where to look closer. But even Deep Throat frustratingly would do no more than provide direction and confirm what the reporters' other sources had only hinted at. "I can't and I won't give you any new names," he told them.

Along the way, Woodward and Bernstein found signs of an effort by the White House to conceal its involvement in the minor burglary of the Democratic headquarters at the Watergate Hotel. That effort grew into a conspiracy with far-reaching influence.

The investigative work of Woodward and Bernstein began to uncover a scheme of clandestine political activity that was leading, ever more clearly, to the US Attorney General, John Mitchell.

Woodward looked through his contacts for a particular link. He wanted someone who had contact with the more suspicious names he was following up on. He identified an attorney at the Justice Department who was getting fed up with the whole cover-up swirling around in the highest levels of government. Upon making contact, the angry attorney revealed more in that phone call than perhaps he should have.

"It was strategy," the man said, "Basic strategy that goes all the way to the top. Higher than him [Mitchell] even."

This was the breakthrough Woodward and Bernstein were seeking. If this man's words were true, they implicated the only person in the United States government higher than Mitchell: the President.

Danger and risk

The *Post* ran Woodward and Bernstein's new story. It was headlined, *Mitchell Controlled Secret GOP Fund* connecting the Attorney General to the slush fund scheme. If Mitchell was involved, then he had to be acting on Nixon's orders.

Journalism is not a field for the faint hearted. If you are going to try to bring down a popular political figure like Nixon who commandingly had won his position in the most recent presidential election, you had better be prepared for the threats. In the 1970s, for Woodward and Bernstein, these threats were delivered in subtle ways.

Woodward and Bernstein's investigative work grew ever more dangerous, with risks to the reporters, to the editors of the *Washington Post*, and to the reputation and validity of the free press in the United States. The more dangerous it became, the more vital the need to establish the rock-solid truth of every fact in every story, for their reputations and the safety of all concerned.

Missteps on a grand scale

Matters came to a head when a story based on testimony given to a grand jury investigation of the secret slush fund was published.

Woodward and Bernstein had, through various confidential sources, identified and confirmed that four out of the five people high up in the White House could be named as co-conspirators to cover up their own and the President's involvement in the Watergate burglary. The final missing name was that of H. R. Haldeman, Nixon's Chief of Staff. Haldeman controlled the Nixon campaign's secret fund for political espionage and sabotage.

Ben Bradlee, the *Post's* editor, would not let the story run unless they could find four sources for the story. The stakes were too high. The story was accusing the very highest levels of government of conspiracy. To Woodward and Bernstein's frustration, they only had three sources, and Deep Throat refused to be named as the fourth.

What if their assumptions were wrong? What if, in their pursuit of the "truth" as they saw it, they were inflicting permanent damage not only on President Nixon, but on the institution of the presidency itself?

The truth of the Watergate break-in and cover-up story, if it ran, had to be unassailable. They needed that fourth source.

With the daily deadline pressing, they sent the story to the composing room to be set in linotype, ready to move a reporter's typewritten story into a format ready to be printed in the physical newspaper. The linotype was set with a space after to include a confirmation or denial, should one be issued by the White House, after the customary preliminary reading.

The White House denied any wrongdoing.

Their denial was inserted and ran with the story the next day, October 25, 1972, under the headline, *Testimony Ties Top Nixon Aide to Secret Fund*.

Immediately, one of the key sources that Woodward and Bernstein had depended upon denied he had ever named Haldeman. Either that person was lying, or Woodward and Bernstein had made a grievous mistake. The *Post* had to defend the story or issue a retraction.

Bradlee delivered a brief statement: "We stand by our story."

The man they called Deep Throat told Woodward that their story had committed a royal screw-up.

But it was not the end of the Watergate investigation, not by any means.

Chapter 22
Investigative Reporting

The Truth Emerges

Woodward and Bernstein published additional stories on the secret fund and other, more minor, instances of collusion high up in the White House. All the while, they chased down leads looking for that crucial link to Haldeman, Nixon's Chief of Staff.

The White House's denials were in vain. The story had become too big to contain.

By this time, other news organizations, including *Time Magazine* and the *New York Times,* had their own investigative teams assigned to it. John Dean, White House counsel since June 1970, began cooperating with the Senate's Watergate investigators, while he was still working for Nixon. He was fired from that job on April 30, 1973, the same day the resignations of Haldeman and Ehrlichman (Nixon's domestic affairs advisor) were publicly announced. Dean ultimately testified in a full Senate hearing.

The facts led conclusively to President Nixon himself, who had been actively involved in the growing conspiracy to contain the story. Nixon's habit of recording confidential conversations in the Oval Office was revealed, and the content of those tapes led to his downfall.

Rather than subject the nation and himself to the painful ordeal of impeachment, President Richard Nixon resigned the office on August 8, 1974.

Woodward and Bernstein's tireless teamwork revealed the full extent of the criminality and coverup. It forced the resignation of a lawless United States President who only two years before had won his position in a decisive landslide election. Woodward and Bernstein's determination, the essence of good investigative reporting, became a model for all such journalism in the decades to follow.

A work of staggering importance

The goal of Woodward and Bernstein's investigation had never been to overturn the presidency. It was simply to get at the truth. That was their commitment, and the Washington Post's.

Had they not both been assigned to the same story at the beginning, Woodward and Bernstein would likely not have come together as a team at all. Although they went their separate ways professionally, the two remain friends fifty years on, talking with each other frequently and still sometimes arguing.

We know their story best from the book they wrote about their work on the Watergate investigation, *All the President's Men*. It is superbly written, in the style of a tense political thriller. Their book was an instant bestseller in 1974, and was made into a highly popular movie released in 1976. It starred two of the greatest actors of the era: Robert Redford played Woodward, and Dustin Hoffman was Bernstein.

These, along with *The Final Days*, their follow-up book about the fall of the Nixon administration, brought the art of investigative reporting into public awareness. Woodward and Bernstein did not invent good investigative journalism, but their work was so much in the public eye that for perhaps the first time people in the United States had a better sense of what it was, and how it could be used to try to keep a government honest.

It is a cultural phenomenon of immense importance.

With the political and legal issues arising from the January 6, 2021 insurrection at the US Capitol, Woodward and Bernstein have again become popular, go-to guests of television news shows and podcasts.

They provide useful insights in the abuse of power, comparing the Trump era with the Nixon era, and the lessons that can be learned from both.

The lessons from their work are vital to a democracy and it's important that we all carry them forward. A democratic society depends upon having a free press to shine light into the darker inner recesses of the political machinery of government, where powerful forces can seek to retain power and subvert the will of the people.

Thanks to the pioneering work of Woodward and Bernstein showing how investigative journalism should be done, it is much harder today for elected officials and their administrations to get away with criminal behavior – not only in the United States, but in other nations around the world who value a free press.

Nixon tried and failed. Others will try in the future.

The masthead of today's *Washington Post* says, "Democracy dies in darkness." It was placed on the front page during this century, but its roots come from a very small team: Woodward and Bernstein.

★

Mastheads appear on newspapers. They are also the part of a ship. The next small team, the captain and navigator of a clipper ship – a husband and wife team in the mid 1800s – used creativity for different purposes: to sail across the seas.

In an era when Susan B. Anthony and Elizabeth Cady Stanton were just beginning their fight for women's rights, Josiah and Eleanor Creesy were sailing a clipper ship. What was unusual about that was Eleanor was the ship's navigator. She set a record that would stand for over 130 years. Their story is next.

The *Flying Cloud* as depicted in a 1917 painting by Antonio Jacobsen.

Josiah And Eleanor Creesy

A sailing vessel is alive in a way that no ship with
mechanical power will ever be."

—*Aubrey de Sélincourt*

I n 1851, the quickest way to move people and manufactured goods
from New York to San Francisco was to load them onto a ship
and then sail from the east to the west coast of the United States
around Cape Horn at the southern tip of South America. The
transcontinental railroad had yet to be built; it would be another
sixty years before the Panama Canal linked the Atlantic and Pacific
Oceans.

The treacherous, 13,000 miles (20,900 km) sea voyage could
take three-and-a-half months or more to complete. Hard sailing
was needed, driven by a brave and skillful sea captain, on a course
directed by the ship's navigator. Many ships were lost to the
challenging seas around South America, foundered by the adverse
winds, gigantic waves, and vicious storms of the notorious Drake
Passage. Even the best pairings of a sea captain and navigator would
face challenges in the Drake Passage.

It was around this time that the three-masted clipper ship emerged as a faster merchant sailing vessel. Long and lean, this new design also became the most desirable passenger vessel, with accommodation for a few well-paying travellers on the deck.

The Clipper Ship

"Clipper ship" calls to mind the image of a graceful three-masted ship powering her way across the seas, propelled by a cloud of full-bellied white canvas sails, and with a strong bow parting the waves to form a white wake. The sun shines benignly on the scene, full of promise of faraway exotic ports.

A romantic image, surely, and not entirely inaccurate.

The sailing ships of the 1850s did, from time to time, display this image. And this was the experience of those lucky enough to be sailing on the wide and empty oceans, on a sunny day, under the balmy influence of steady trade winds. However, such circumstances were relatively rare.

A passage from New York to San Francisco back then spent less than a fifth of its three-plus months making easy headway southward, off the east coast of South America.

For the remaining four-fifths, the navigator had to make mathematically detailed reckonings of the ship's constantly shifting location on the trackless seas. He needed to have an astute understanding of climate and weather, knowledge about the mechanics of the rig, sail and hull, as well as have an ability to work with the crew.

Taking latitude readings by sextant and working out the mathematics of longitude, complicated though they were, was the easiest or most predictable part. As for the square-sailed rigging, the masts with their yards crossed far above the sloping deck, and the shape and condition of the wooden hull – these at least were inanimate, and subject to the immutable laws of gravity, and the physics of tension and compression.

But what about the crew?

They were mostly a motley gang of otherwise indigent men (at

Josiah and Eleanor Creesy | 171

this time, women would not have been taken on as crew members). Some had sailing experience; others did not. Their shoreside social life centered in each port's sailor town of bars and brothels. Once on the ship, their lives were abruptly changed to one of sobriety and male-only company. And they were not always the most congenial people.

Many of the sailing-ship crew knew no other life than to spend a few short days carousing on shore until the money from their last voyage was all spent. Often still drunk from their last night out, they tossed their sea-bags on deck, signed their name on the ship's articles – or made an X if they had not yet learned to write – and chose one of the wooden bunks in the fo'c'sle to sober up on. Following this rough introduction to their new shipmates, the sailors were divided into one of the two watches, to stand to duty four hours on, four hours off, for the duration of the voyage.

Having left one harbor, the sailors would not see land again for three, four or five months if they were lucky. Or if they were very unlucky, never again. A ship would not be posted "missing" until at least six months had passed after failing to arrive at her port of destination.

Would any clipper ship captain want to subject his wife to this?

Josiah Perkins Creesy did. And in 1851 Eleanor Creesy, serving as the ship's navigator, piloted the brand-new clipper ship *Flying Cloud* to a record-setting maiden voyage from New York to San Francisco.

Captain and Mrs. Creesy were a formidable team.

Origins of the Team

Josiah and Eleanor were well suited from the moment they met at a dance. They were both 27 years old and neither had a desire to have children.

Josiah Perkins Creesy was a sailing ship captain long before he met Eleanor. He was already well-versed in the complications of celestial navigation. No man could claim to be a captain without having such skills. Only a fool would depend entirely on someone else's calculations, and the owners of the ships on which he was commissioned to command had no use for fools.

Although every ship could have only one captain, every ship must have more than one navigator. Most often the first mate had charge of this task. The more minor second, third, and fourth mates, had to have some navigational skills of their own if they were serious about moving up the ranks.

Captain Creesy did not need a navigator. He was one. But he did need a wife.

He found in Eleanor a woman who was willing to travel the ocean seas with him, and to share the relative comfort of the captain's cabin on the ships he would command. She brought a calming presence to the male-dominated crew.

She was born into a maritime family in 1814 and grew up on the water. Her father owned and sailed the schooner *Californian* out of Marblehead, Massachusetts. Her parents doted on her, making sure she had a good education, one that was fitting in a society where a young woman's only prospects were marriage and family.

But Eleanor also had a maritime education. Her father would not get the son he really wanted for another 23 years, so he treated his only child as one, even though she had been born a girl. He wished he could raise her to take over the family business, but the times being what they were, it was not likely she would ever be in a position to do so. So, he did the next best thing.

He brought her on his sea voyages up and down the coast and, at her request, taught her the complex skills of navigating a ship. Not just the rudiments, but the fine art and practice of celestial navigation.

Eleanor had a hands-on knowledge of the ways of the sea, and a desire to share those ways with a prosperous captain, should one pledge his troth to her.

In time, she became bored with shoreside life and the prospect of entering the dreaded "spinsterhood", to which every white woman of her era was condemned to if she entered her thirties unmarried. Marriage to a successful sea captain would be a good option for a woman who wanted to see the world, and when Josiah made his offer, she did not hesitate to accept.

Captain and Mrs. Creesy tied the knot in 1841, and for the next ten years sailed together on the ship *Oneida* from New York to China and back, bringing crates of Chinese tea to the American market. With no children to tie her down, Eleanor had the freedom to travel with her husband on board his ship, wherever and whenever the tea trade sent him. Their life on board suited the two of them perfectly.

Becoming a full-fledged navigator

During the ten years Eleanor navigated the *Oneida* out of New York, around southern Africa, through the storm-ridden "roaring forties[12]," through the Sunda Strait, and onwards to Borneo, the Philippine Sea, and then reaching the tea ports of China. And back again.

Eleanor did much more than take the occasional sun sight with the sextant; she handled the entire range of course-plotting for those five voyages. Josiah and his first mate managed the crew's work on the ship, setting sail and taking in canvas as needed to keep to the course, and directing repairs when storms damaged the rig. Josiah and Eleanor made a perfect team.

To truly understand the magnitude of being an exceptional navigator in that era, we need to look beyond the mathematical computing of degrees North or South latitude, or East or West of the Greenwich Meridian.

Think first of how the world must look from the deck of a ship at sea, on any of the seas of the world. Barring the incidental cloud or crest of wave, the view is pretty much the same wherever the ship is. Without GPS, how can any point on the wide and trackless sea be differentiated from another? How do we know which direction to go to reach shore while bypassing the large and small land masses, islands, rocks, and shoals that lie between the points of our departure and our destination?

Picture a map of the world showing the oceans, the continents and the equator all on one page. Mark your ports of call for your voyage

---

[12] A latitude measure. Between 40° and 50° South there are intense and persistent winds.

and draw a line from one to the next. In the case of our tale, from New York to San Francisco, you'll find your route is compelled to go around the base of South America.

Such voyages must cross the equator, so one of the first things to consider and confirm is how far north or south of the equator the ship is on any given day. No less important is the ship's location east or west of an imaginary line from pole to pole that's been fixed at Greenwich, England. These mathematical coordinates were the only way to define the position of the ship when out of sight of land. For a clipper ship going around "the Horn," there would be no land in view – ever – from the time she left New York until she entered the bay at Golden Gate in San Francisco.

Despite the empty horizon of the sea, there are ways to precisely find one's exact location on the globe by using degrees north or south and degrees east or west. Such precision demands a sophisticated level of training in the use of complicated measuring devices, and a well-rounded background in mathematics.

It required an analytical mind, a strong disposition, an attention to detail and excellent training. Eleanor had all this and more.

Defying Convention

Although Josiah had no real need for another navigator, as he could handle that task himself, Eleanor wanted a more rewarding position on the ship than merely being the captain's wife. It required both to look beyond the traditional (and rare) role for women as helpmate to the captain and nursemaid to the men on board.

Eleanor could navigate the ship's position through all the seas of the world. And Josiah would remain the captain, the commander of everything and everyone on the ship. That was his legal position. But Josiah and Eleanor figured out how to join forces in a meaningful way that transcended the traditional distribution of power within a marriage. Their new-styled union was first put to the test on the voyages of the *Oneida* from New York to Canton (now called Guangzhou), in southern China.

The *Oneida*

There is little real data to clarify exactly how the decision to work together came to be. Had Josiah relented to his bride's persuasive argument that she should have the responsibility for the ship's navigation?

One might presume that their first trip to China was good practice for her. It was a chance to turn her US coastal experience into something more refined, in a more global and remote setting. The use of the instruments and the mathematics were as she had learned them. But there was nothing quite like a months-long lesson in daily practice to harden that theoretical knowledge. Eleanor kept a daily round of notations in the ship's logbook, proof for the owners and the world to see that she could do this.

However, the owners needed convincing.

Given that era, even Josiah most likely had his doubts at first. But the ship's logbook, with its daily notations proving the course taken made for a smooth and relatively fast round trip, should be sufficient to convince any businessman or captain that Eleanor Creesy had the requisite skill. And from that voyage onward, no one questioned her ability to deliver a ship safely and on time to whatever destination could generate the most generous profits.

She was not, of course, the captain.

She could guide the ship through the trackless waters, but she had neither the legal right nor the force-of-will that commanded all life on deck between the ports. That belonged only to the captain.

On board the ship, the captain's word is law. He is dictator of his little sea-borne world. He alone decides what must be done, who should do it, and how it should be done. And if any sailor raises an objection to fulfilling his duty, that sailor will be guilty of mutiny. And the penalty for mutiny, enforced on shore by land-bound courts when the ship comes into port, is in most cases, death. Such objection is not undertaken lightly by anyone who has signed his name in agreement to the ship's articles.

On most ships at sea in 1851, the captain is a lonely man. He lives alone on board, and most often dines alone as well. He seldom takes counsel from anyone, even his first mate. He will hear no argument and accept no defence from any recalcitrant seaman.

The seamen are, as a class, a very tough lot. Their pay is low, their rights few, their work constant, their food poor. Their bunks in the foc'sle are a hard board with only a thin straw tick for comfort. On many ships, their worst impulses are held in check only by the threat of immediate and violent corporal punishment by the first mate. The captain seldom speaks to the men of his crew directly – he tells the first mate what he wishes to have done, and the first mate sees to it that the job is accomplished.

This was the way on all tall ships in 1851. Seamen's rights wouldn't be protected by labor law for another half century at least. In the meantime, every vessel at sea was a hotbed of barely controlled working-class aggression, held in check only by the force of the captain's will.

It is small wonder that few captains invited their wives to come along, and even fewer wives took them up on the offer.

The *Flying Cloud*

Clipper ships were all the rage in those days, and were very much in the news in cities on the eastern seaboard of the United States. A week could not pass without one of the daily newspapers breathlessly touting the launch of one of these "greyhounds of the seas."

Longer, leaner, and faster than the full-bodied wooden sailing ships of the past generation, they looked like the thoroughbred racers of the shipping trade.

Even though clipper ships could beat their predecessors by a knot or two, this was not enough to matter over the course of a months-long passage from one port to the next, since almost every voyage involved battling the vagaries of ocean storms and variable winds. Success of the voyage came down to the combined skill of the captain and the navigator.

Most often, in all the sailing ships and steam-powered packets, the captain was also the navigator. The skills of taking stock of the ship's position by sun sights and commanding that ship were mutually interdependent. No shipping company would hire a man for captain who could not deliver on both counts. But for Grinnell, Minturn & Co., the makers of the newly launched *Flying Cloud*, they had a double stroke of luck in the Creesys – a commanding captain and a highly skilled navigator.

Clipper ships were built for hauling cargo, which was where the profits were in the shipping trade. Although earlier types had no accommodation for passengers, the *Flying Cloud* and her sister ships made a modest attempt at providing such for travelers: a cabin on deck with small, unheated staterooms and the barest minimum of built-in furniture. There was no great cabin or saloon. In fine weather the passengers lounged about on deck, trying to stay clear of the sailors moving about the ship tending to the sails; in foul, they sequestered in their rooms.

But in 1851, the California gold rush was in full swing. There was no good or fast way for anyone to get from the eastern seaboard of the United States to San Francisco's Golden Gate. For those who could afford the premium rates, a ride on a clipper ship around Cape Horn was as good a way to travel as any of the other cross-country routes. And, for those in a hurry to strike it rich in the goldfields, speed – or the illusion of it – was worth the cost.

When the *Flying Cloud* set sail from New York on June 2, 1851, her holds were full of general cargo to supply the shelves of merchants and households in the rapidly growing city of San Francisco. Only two years earlier, it had been a sleepy town on the other side of the continent, but now it was the destination for thousands of men seeking their chance at riches.

On deck, eleven passengers occupied the six staterooms. No one knew on that day that they would be sailing into history. Not even the captain and his wife, although they thought, with their new ship and a new set of sailing directions, it was a possibility.

New thinking confronts the old ways

For all the risks sea captains took in hauling cargo over the stormy seas all around the world, they were a conservative lot. They tended to follow the same tried-and-true sailing directions that had been the rule of thumb for centuries.

When a ship set sail from New York with the aim of going around Cape Horn at the southern tip of South America, traditional navigation mandated that the ship must first go well to the East – a thousand miles or so toward Europe – before heading South. This course was supposed to avoid adverse winds and currents that were believed at the time to confound an easy passage around Cape São

Roque, the easternmost point of Brazil.

The route was so deeply set in their minds that almost no captain or ship's navigator dared to challenge it.

Matthew Fontaine Maury was head of the Depot of Charts and Instruments at the US Naval Observatory. As an oceanographer, he realized that if he accumulated the details in the logbooks of the many ocean-going ships, he would have a wealth of observations about wind and weather on any given day, for any square mile of ocean on Earth.

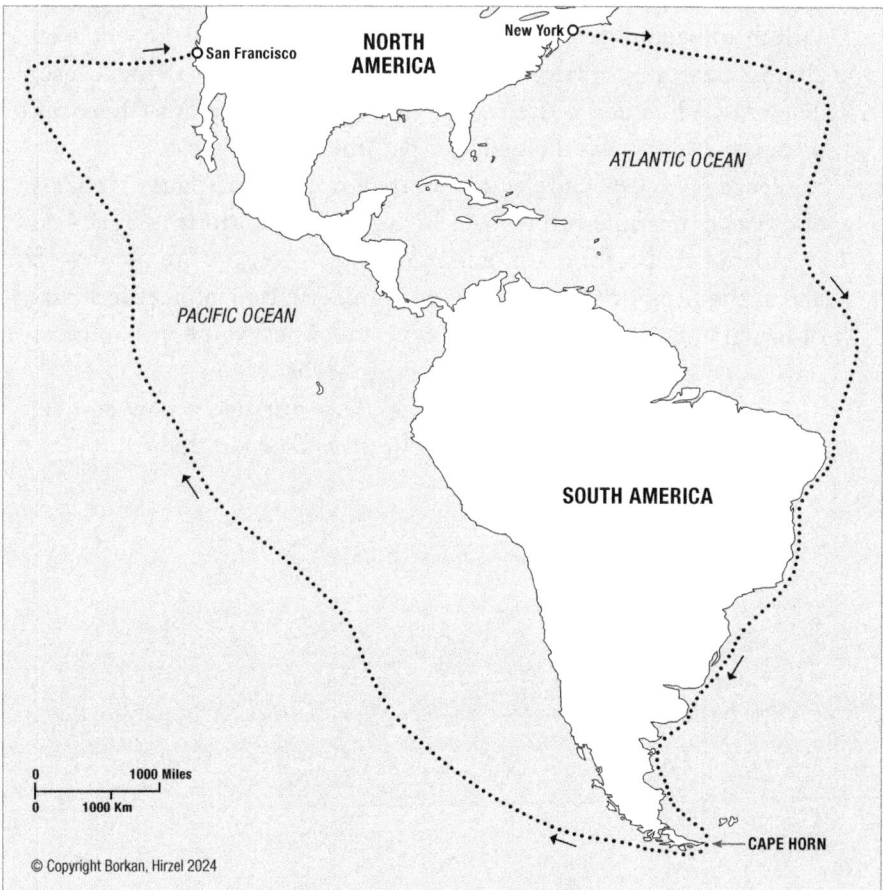

Approximate route of the Flying Cloud.

Written down in all these logbooks were millions of data points from thousands of voyages. In what is surely the earliest conception of "big data," he assembled all this information into a comprehensible spreadsheet based on the map of the world (long before the invention of Excel). He found answers to the big questions: Where were the most favorable trade winds blowing, and from which direction and at what strength? Where were the limits of the doldrums of no wind? What was the set of the ocean current in any given location?

Maury's charts, first published in 1847 as the Wind and Current Chart of the North Atlantic, held the key to faster journeys, but sailing masters – as stated, a generally conservative lot – biased against new and untried ideas, were hesitant to put them to use. The ones who did reported excellent results, and greatly improved passages. Maury was dubbed the "pathfinder of the seas."

Eleanor Creesy had bought her own copy of Maury's charts and was determined to try out his recommendations. Maury had calculated that sailing a more southeastern route from New York across the Atlantic Ocean, and then coming in tight along the coasts of Brazil and Argentina was the fastest and best route. Josiah must have concurred with Eleanor's desire to try it.

Against all convention the *Flying Cloud* pursued a new course, following the Maury's sailing directions, and began their run.

Chapter 25
Navigation

Around Cape Horn

When they reached the doldrums in the mid-Atlantic, they crossed that area of no wind near the equator in a matter of four days; those captains who declined Maury's suggestions were two weeks getting through. Better winds favored the *Flying Cloud* south along the coast of Argentina as well.

Of course, no book of generalized sailing directions can guarantee the presence or absence, or direction of wind on any given day. While it is the navigator's task to locate the ship on the globe, the captain's is to use the winds as they are found, to direct the ship as nearly as possible in the course indicated. Detours in search of the most favorable wind are a part of the process.

Dangers abound

Squalls, gales, storms and lightning strikes occur at their own rate, regardless of what the climatic conditions may indicate. When a furious gale descended on July 9, 1851 and prevented any latitude observations on the following day, Josiah shortened sail to carry only close-reefed topsails and staysails, but this was not enough.

A burst of wind ripped the staysails to tatters, and both the fore and main topmasts split and carried away, tumbling into the sea. Rolling uncontrollably, the ship dipped her yardarm ends into the sea on either hand.

Any captain worth his salt is prepared to deal with emergencies like this at sea.

Shouting through his speaking-trumpet to be heard above the wind, Josiah Creesy gave the orders, and his crew responded with alacrity, to get the ship under control. When the gale eventually died down, he had the deck cleared of the mess of splintered timbers and tumbled gear, and set the crew to restoring the rig.

They were drawing nearer to Cape Horn, the southernmost tip of South America, home of the most fearful storms and adverse winds to be found on the planet.

The wind drove the spume off the wave tops in horizontal sheets, where it piled up on the sides of the deckhouses. The shortest course lay dead ahead, through the Le Maire Strait. The ship was another two days getting a straight shot into the strait. Many ships avoided this narrow alley of seaway, where once the ship is committed there is no turning back. Many ships stuck to the relative safety of a time-consuming alternative: outside and around the island.

But the Creesys trusted the navigation and Cape Horn lay but a few more storm-tossed miles south.

Maury's new sailing directions provided no alternative to facing the dead-on headwinds blocking their way west. Here is where luck rather than the skill of the navigator and the competence of the captain comes into play.

The winds in this region, the infamous Drake Passage, sweep around the world from west to east, unrestricted by any land mass between Cape Horn to the north and the Antarctic Peninsula to the south. The place has the melancholy designation "the graveyard of ships," where over eight hundred have been overwhelmed and sunk without a trace. Only on the rarest of occasions do the winds die down to allow an easy passage. Most ships of that era took weeks, sometimes months, to make this passage from the Atlantic into the Pacific Ocean. Captain Bligh in the *Bounty* gave up altogether, deciding it would be easier to take the longer route around the world to his destination, Tahiti.

Luck favored the *Flying Cloud*.

The storm had brought that rarest of winds, an easterly fresh breeze to propel her easily in the right direction, toward the west and the Pacific Ocean. In only four days the Creesys had rounded South America and were on a new course, north to San Francisco.

A fast passage north

The ship was at last headed north through the Pacific Ocean, but the work was far from done. She would still have to contend with the variable winds and doldrums that interrupted the broad belts of southeast and northeast trade winds along their course. The best ways to make use of them were already well anticipated by Maury's sailing directions.

The book and maps he created indicated a course swinging far to the west – a greater distance over the water but a more efficient and, therefore, shorter time in transit. To call this passage north a "romp", as some have done, is to diminish the ongoing work to the ship and its crew, but compared to the outward-bound passage from New York, it was.

No damage was accrued from storms, and no adverse winds, only the standard shipboard routine under normal oceanic weather.

Comparing her sun sights from noon July 30 and noon July 31, Eleanor found that the ship had covered a truly remarkable distance – 374 nautical miles in 24 hours, a world record for any sailing ship.

The captain noted the achievement in his log, along with all the other details of the day's sailing – the setting and taking in of sail as squalls hit. But by six o'clock, he ordered sails to be taken in as the storm grew in intensity, and the ship began to labor under the strain. But still the ship drove on, and by August 2, she logged another 992 nautical miles, or 1,440 statute miles in three days.

Once through these doldrums on August 19, the way was clear for a record-setting voyage. Well off the coast of California, the time had come to take advantage of favorable winds and by the 24th, the *Flying Cloud* was heading straight for the Golden Gate. It should only be a matter of days now, but baffling winds in the variables slowed the ship's progress to a mere 52 miles (84 km) per day. The captain maintained all sails set, ready to catch the wind. By two in the morning of August 31, the ship lay hove to just 30 miles (48 km) off the Farallon Islands, awaiting daylight before the last few hours run through the narrow channel known as the Golden Gate, with the burgeoning city just beyond.

The *Flying Cloud* splashed her anchor at San Francisco's North Beach that day. The run from New York had set a new record, 89 days 21 hours, better than the previous record by over six days.

Maury had estimated a theoretical (perfect combination of winds and currents in the best sailing months of the year) passage of 85 days. *Flying Cloud* sailed in the most adverse time of the year for Cape Horn (arriving there in winter) and during its voyage was twice partially dismasted.

Eleanor and Josiah Creesy, navigator and captain, could claim this sailing record for only four years.

They broke it a second time by 13 hours in the same ship, along the same course in 1854. And their new record would stand challenged but unbroken for 135 years.

Legacy

Records aside, the challenge for any team is to combine their talents and ambitions, to achieve something remarkable. In the case of the Creesys, they aspired to improve the world's commerce by making faster and safer deliveries. Captain and Eleanor Creesy, aided by their inspired choice to use Maury's charts, proved that efficient sea travel could be made between far-separated ports.

Their success was also thanks to a combination of beliefs.

Eleanor's remarkable navigational skill – a rarity for a woman to have such training in that era – and Josiah's belief in his wife as a ship's navigator enabled their success. Skill and trust are two of the many ingredients that transform a team into a dynamo. If one or both are lacking, the team will be unlikely to achieve their potential.

As for the Creesys, their lessons, well-taken by those who cared to study them, remained a guiding light for sailing ships. In 1914, the Panama Canal opened, rendering the difficult sea lanes around Cape Horn a thing of the past.

The Creesys' record stood until 1989 when it was surpassed by a team of three, Warren Luhrs, Lars Bergstrom and Courtney Hazelton sailing a high-performance racing sloop. Their ship, *Thursday's Child,* carried no cargo or paying passengers. It did have something that the *Flying Cloud* did not – a fabricated hull made with space-age materials.

★

Now we move on to very small teams that are much better known than the Creesys. These are real and fictional duos and trios that star in Hollywood films and television shows.

The iconic Hollywood sign. *(Photographer: Carol M. Highsmith)*
(Library of Congress)

Chapter 26
Hollywood and Popular Culture

Pre-War Era

"Louise, no matter what happens, I'm glad I came with you."

—*Thelma, in the final chase scene in the film,*
Thelma and Louise.

Wherever you look, whether it is in Hollywood movies or on stage, in comedy or drama television shows, children's cartoons, comic books, magic shows, and music recordings and festivals, you will find very small teams of highly entertaining and talented people.

In whatever era you grew up in there were pairings of people who created exceptional and memorable work that changed the entire genre in which they appeared. There were double acts to entertain, thrill and amuse you and your generation, whichever one it may be – baby boomer, Gen X, Millennial, Gen Z, or even pre-baby boomer – and their influence continues to this day.

Pre-war comedy

Before the advent of "talkie" motion pictures, vaudeville acts—live onstage comedy entertainment—could be found in communities throughout the English-speaking world. With the advent of broadcast radio and then television, physical comedy teams like the Marx brothers, Laurel and Hardy, the Three Stooges, Abbott and Costello, and the female duo Thelma Todd and ZaSu Pitts emerged.

Each pairing had a unique style of comedy that relied on one of the two people to play a 'straight' person, or the non-funny one, who reacted to the other's comedic actions. While it might be expected that pay would have been split equally between them, the person playing the 'straight' one was, in some groups, paid more than their comedic partner, because it was deemed harder to play the non-funny character.

As mass media developed, evolving from "talking" motion pictures and radio in the 1930s into television in the 1950s and onward, these teams adapted to the new opportunities to reach ever-wider audiences. Their careers spanned decades, and they thrilled audiences with their wit, slapstick comedy and impeccable comedic timing.

Many of their routines were preserved and distributed in brief 15 to 25 minute, two-reel comedy short films for distribution in movie houses around the country. These found their way into the early days of broadcast television in the 1950s, and from there into our collective consciousness today.

What is considered to be one of the best comedy routines of all time was Abbott and Costello's *"Who's On First?"* It was first broadcast on radio in 1938. While there may be some discussion as to whether they originally wrote the script, Abbott and Costello perfected its rapid-fire delivery to reach an astonishing new level of comedy.

It requires some understanding of American baseball terminology and relies on word play. In the skit, "Who" is the name of the baseball player covering first base, and also the question Costello wants answered, "Who's on first?" (baseball jargon meaning, "who is the first baseman?")

Straight-man Abbott replies, "I'm telling you. Who is on first." This creates ever-increasing confusion for Costello.

Costello tries to clarify by asking Abbott, "If you are paying the first baseman, then *who* picks up the money?" assuming he'd be told the player's real name. Abbott replies the ballplayer gets the paycheck but sometimes the ballplayer's wife picks it up.

Costello, more confused, asks, "Whose wife?" To which Abbott replies, "That's correct. *Who's* wife!"

Costello is none the wiser, and the comedy routine accelerates from there through the six-minute sequence, which also features players named "What" on second base and "I Don't Know" on third.

The tightly scripted routine was perfectly acted – an endearing triumph that could only have been achieved by one of the greatest small comedic teams in history. It is comedy gold.

Perfect pairings were to be found in other genres of popular entertainment.

Fred Astaire and Ginger Rogers danced their way into the collective hearts of their generation. The music and lyrics of one of their ten movies, *Shall We Dance,* were composed by another very small team: George and Ira Gershwin.

Comedy double acts were also popular in the United Kingdom. The most loved and revered of them was the duo Eric Morecambe and Ernie Wise, simply known as Morecambe and Wise. They started working together on radio in 1941. After World War II, they became even more popular in film and later on television, combining song and dance routines with comedy skits.

Many of their lines and catchphrases went immediately into common British usage. One example is when an ambulance speeds by, even to this day someone might exclaim, "He won't sell many ice creams going at that speed, will he?"

Morecambe and Wise's long career influenced British humor throughout the post-war years and continues to influence it today. In their heyday, their Christmas television specials in the 1970s were seen by up to 28 million viewers, which was almost half of the UK population at the time.

Batman and Robin

As mass media grew in power, the prevalence of duos and trios became significant in print and literature as well.

Batman was first written into a comic book in May 1939, and his sidekick Robin was added in April 1940. From then on, they were an inseparable pair, protecting Gotham City from villains like Joker, Two Face, Penguin and Riddler, except of course when one or the other was captured in one of their many escapades.

Batman and Robin were brought to life by another powerhouse duo. Bill Finger wrote the early stories and Bob Kane illustrated them. Not all teams, however, have happy outcomes.

For creating one of the greatest, best known, most lucrative superheroes in history, Bill Finger received little, if any credit and almost no money. He died poor and anonymous, just before his 60th birthday. Bob Kane on the other hand got the byline credits, enormous financial rewards and kudos as the "originator". Author Marc Nobleman discovered that not only was it a team effort, Bill Finger heavily influenced the design, including the characters, as well as Batman's costume, Batmobile, Batcave, bat symbol, his origin story, and the cast of colorful and memorable villains.

At around the same time that Robin and Batman came together, another action duo appeared in film and television, but these two were not companions. They were mortal enemies. The television cartoon series *Tom and Jerry* featured a vicious house cat named Tom and an even more vicious and retaliatory home invader – a mouse named Jerry.

Tom and Jerry would fight each other endlessly in entertaining chase sequences, but never seemed to inflict permanent damage, despite using lethal implements including axes, guns, knives, and sticks of dynamite. On occasion they'd put their differences aside and team together to fight an external force.

★

But vaudeville, comedy, comic book and cartoon characters were only the beginning of the entertainment outlets available to very small teams.

1950s and Beyond

I n the late 1950s new teams appeared, especially in cartoons like *the Flintstones*, which featured two pairings: friends Fred Flintstone and Barney Rubble, and their long-suffering wives, Wilma and Betty. The cartoon women's friendship was formed out of their combined exasperation over their husbands' antics.

Rocky and Bullwinkle was another cartoon pairing. Rocky (a squirrel) and Bullwinkle (a moose) were a small team. Their on-going mission was preventing Boris and Natasha, an evil Russian pair of secret agents, from conducting their dastardly deeds. The episodes were filled with pranks and mis-steps by one or more of the teammates.

Lucille Ball and Desi Arnaz were a real-life pairing: a husband-and-wife comedy team, and business partners. They made only two feature films together, *Too Many Girls* in 1940 and *The Long Long Trailer* in 1954. They leveraged those successes into one of the most successful and best-loved television series, *I Love Lucy* (1951-1957). Their production company Desilu introduced camera and staging techniques still in use today, and had a lasting effect on popular television entertainment.

Audiences responded well to this sort of interactive, light-hearted teamwork. It was something they could see in their own relationships, where humor and good-natured banter could help relieve the tensions of work, marriage and families. Production companies seeing its fundamental value, expanded the concept into larger, more complex cinematic works.

Hollywood: good versus evil

By the 1960s, Hollywood movies began featuring very small teams, some of whom were outlaws but portrayed in a sympathetic light. These films challenged the convention that there are people who are only good or only evil.

Bonnie and Clyde came out in 1967, based on the true story of a 1930s couple who robbed banks in states like Texas and Oklahoma. The actors were Warren Beatty and Faye Dunaway. This film transformed how audiences perceived lawbreakers, as it was unclear whether viewers were supposed to cheer for the police or the villains.

Butch Cassidy and the Sundance Kid, also based on a true story about a pair of train robbers on the run from the law, premiered in 1969. Like *Bonnie and Clyde,* this movie also featured a pair of much-loved, acclaimed actors of the era: Paul Newman and Robert Redford. Newman and Redford's camaraderie was so popular, that they teamed up to produce two more movies, *The Sting* and *The Sting II.*

What was remarkable was how important the teams were to these two films and how they captivated all who saw them. These were fictional accounting of real teams that gave the public a vicarious view into their interpersonal dynamics. The audiences experienced the emotions that the characters had for one another.

All of this set the stage for the breakthrough movie of the early 1990s, *Thelma and Louise.*

In *Thelma and Louise,* the fictional title characters played by Geena Davis and Susan Sarandon, were not villains until circumstances turned them into a pair on the run from the FBI. It was a powerhouse performance featuring two women taking control of their lives.

Spoiler alert for the following paragraph: The ending of the film is shocking, and unforgettable. During filming, an alternative ending was proposed: That Thelma pushes Louise out of the car before it goes over the cliff. But Susan Sarandon, who played Louise, fought to keep the ending as it had been originally written – that Thelma and Louise stay in the car right till the end. They were in it together and they would drive off the cliff together. They would stay a team to the end. They kissed, held hands, and chose to "keep on going," as a team.

All three of these movies *Bonnie and Clyde, Butch Cassidy and the Sundance Kid,* and *Thelma and Louise* were vastly popular and each were transformational in how subsequent Hollywood films would portray pairs of people.

Kirk and Spock, Han Solo and Chewbacca

Television also played a key role in showcasing the power of very small teams. *Star Trek* was first aired in 1966 featuring the human character Captain Kirk and Mr. Spock, the non-human Vulcan who could only interpret the outer space worlds they encountered scientifically, but not emotionally. Although radically different from one another, Kirk and Spock (played by William Shatner and Leonard Nimoy) seemed to bring out the best in each other, even when at times they were completely baffled trying to understand each other's reaction to events or risks. *Star Trek* was made into a Hollywood movie (the first of six) in 1977.

The outer-space movie, *Star Wars,* released that same year featured another unlikely but powerful multi-galaxy pair: Han Solo and Chewbacca. Han Solo, played by Harrison Ford, was human; Chewbacca was a Wookiee – a large, lovable beast. In the eleven *Star Wars* movies that were released over the next 42 years, these fictional characters worked together to help the Rebel forces restore freedom to other worlds.

Eighteen years after those first releases, the movie *Toy Story* introduced a new pair. Buzz Lightyear was a plastic Space Ranger toy. Woody was a cowboy pull-string, talking doll with human qualities. Their partnership and friendship were not unlike those of Captain Kirk and Mr. Spock, and Han Solo and Chewbacca. Unlike their precursors, Buzz and Woody's mission was not to save universes but to team up to save Andy, the little boy who owned them as toys.

Comedy and crime fighting

By the 1970s, not all very small teams on television and cinema were cartoons, real life villains, toys or space heroes. In *The Odd Couple*, a comedy television show based on a Neil Simon play with the same name, Felix Unger and Oscar Madison were an unusual pair of divorcees, sharing a New York City apartment. Felix was neat, tidy and precise. Oscar was the complete opposite. For people of a certain generation, the names Felix and Oscar evoke great sentiment.

In the 70s, UK audiences were entertained with the risqué, bawdy and rude comedy of Ronnie Barker and Ronnie Corbett in the prime-time Saturday night phenomenon known as the *Two Ronnies*. At a time when the UK population was 56 million people, their viewing numbers were 18.5 million per episode. Their shows would often end with them in drag performing a song and dance routine, including some written by Gilbert and Sullivan.

At almost the same time as the *Two Ronnies* show was winding down, after being broadcast for 16 years, a pair of British women comedians, Dawn French and Jennifer Saunders, emerged as eminent successors. Their television shows were *French and Saunders* and the popular sitcom *Absolutely Fabulous*.

Other pairings were crime-fighting teams, often with a lighthearted, witty touch. *The Man from U.N.C.L.E* brought together actors Robert Vaughn and David McCallum as secret agents Napoleon Solo and Illya Kuryakin working for a secret international counterespionage agency at the height of the Cold War. Their adversary was an evil organization called Thrush.

Originally conceived as a show featuring single 'good guy' (hence the surname Solo), David McCallum's Russian character was so well received in the pilot, the two-man team was formed. The two characters' popularity in the 1960s inspired a wealth of comic books, novels, soundtrack albums, action figures and board games featuring the duo.

Starsky and Hutch was another example. That show launched in 1975 and featured two detectives solving crimes in California.

Other law enforcement television shows with high performing, very small teams included the partnership of Martin Riggs and Roger Murtaugh on *Lethal Weapon* (both a Hollywood movie and a TV show), and the more modern female crime fighting team of *Rizzoli and Isles*. In the latter's case, Jane Rizzoli is the homicide detective and Dr. Maura Isles is chief medical examiner.

Very small teams in films and television

Many of the very small teams featured in the chapters of this book have been glamorized in films and television. These include a movie about Peary and Henson's quest for the North Pole called *Glory and Honor,* and two films about the Wright brothers: *Kitty Hawk: The Wright Brothers Journey of Invention* and *The Winds of Kitty Hawk.*

Ken Burns made a film about the quest for women's rights titled, *Not for Ourselves Alone: The Story of Elizabeth Cady Stanton & Susan B. Anthony.* The BBC turned Cherry-Garrard's book *The Worst Journey in the World,* about the quest for penguin eggs, into a docudrama. Gilbert and Sullivan's partnership was the topic of several films including *Topsy Turvy,* which described their friction while writing *The Mikado.*

Two of the feature films won Oscars in multiple categories: *Apollo 13* and *All the President's Men.* A docudrama about Hillary and Tenzing's climb of Everest was made, called *Beyond the Edge.* Shackleton and Wild featured in many films and docudramas including a TV mini-series called *Shackleton.*

Teaching very small team dynamics

Sesame Street uses very small teams to convey key ideas to children. This is most notably achieved with the puppet characters Bert and Ernie. They often disagree or annoy each other but stay friends throughout, proving to children that friendships aren't always harmed by differences.

In the original pilot episode shown to test groups, Bert and Ernie had a very small role, but after the audiences loved them, they were given more prominence in the show.

★

Very small teams appear in more entertainment varieties including music, Las Vegas acts and sports. These are discussed in the next chapter. In that chapter we also reveal one of the most intriguing and beguiling very small teams ever created.

Music, Vegas, Sports

Singing duos and trios have been performing and entertaining for centuries. There are many to highlight and their names and music will be recognizable to many. Simon and Garfunkel, and Sonny and Cher are two high-profile pairs who created many hit songs, yet had messy public breakups when the partnerships ended.

Other popular duos included Sam and Dave, Ike and Tina Turner and Hall and Oates, though the Turners marriage ended in divorce and Daryl Hall and John Oates are currently in a legal battle.

The most renowned songwriters were often pairs, such as the exceptionally talented John Lennon and Paul McCartney who co-wrote almost all the iconic Beatles' songs, and Mick Jagger and Keith Richards who did the same for the Rolling Stones' hits. The deep and enduring affection Lennon and McCartney had for one another is evident in the TV special: *The Beatles: Get Back.*

There were also the Brill Building New York City songwriting teams whose music shaped the Top-40 radio format of the 1960s. Among them were the teams of Carole King and Gerry Goffin (*Will You Love Me Tomorrow, Take Good Care Of My Baby*), Tommy Boyce and Bobby Hart who wrote songs for the Monkees like *Last Train to Clarksville,* and Barry Mann and Cynthia Weill (*You've Lost That Lovin' Feelin'* recorded by the Righteous Brothers and *On Broadway* recorded by the Drifters). In a way, these songwriting teams were the George and Ira Gershwin of the rock and roll era.

A trio that deserves greater recognition is the collaboration of Lamont Dozier and the two brothers Brian and Eddie Holland. Known as H-D-H or Holland-Dozier-Holland, they wrote many of the Motown hits sung by groups like the Four Tops, Martha and the Vandellas, Marvin Gaye, and the Supremes.

Many of their hits reached Number 1 in the charts, including *Stop in the Name of Love, Heat Wave,* and *You Keep Me Hanging On.*

Las Vegas pairings

On the live stage, two of the highest profile acts in Las Vegas were very small teams. Siegfried and Roy performed together on stage for 13 years with white lions and tigers. Their act as animal trainers and entertainers was one of the showcase hits at The Mirage in Las Vegas. The act shuttered permanently in 2003 when a tiger viciously attacked Roy during a performance. He lived but was severely injured. Roy defended the tiger's actions, saying he had suffered a stroke and the tiger was trying to take him to safety.

During their joint careers, Siegfried and Roy performed 30,000 shows to 50 million people.

Penn and Teller are exceptionally talented magicians performing together since 1975, in Las Vegas and on television. In their act, Penn does all the talking and Teller acts mute. Teller, who is highly articulate, explains in YouTube videos both why he acts mute and the origins of their act. They have been based at the Rio Hotel and Casino since 2001, making them the longest performing act at the same hotel in Las Vegas history. They claim the reason for their strong partnership is it is based on respect for one another.

Sports: Torvill and Dean

Most team sports involve large teams of players. In basketball there are five players on the court per side, baseball: 9 players, soccer: 11 players, rugby (depending on League or Union) 13 or 15 players. All of these exceed our focus for this book on very small teams of two or three people. But there are two sports worth highlighting because the pair involved transformed their entire sport.

Jayne Torvill and Christopher Dean were British ice dancers who won a gold medal in the 1984 Olympics, and in doing so, turned ice dancing into a worldwide phenomenon. Not romantically involved with each other, they had been winning medals and championships in the lead-up to the Olympics.

Olympic ice dance music has to be four minutes long, but contestants are given a ten-second grace period. The piece of music they chose was the 17-minute Ravel's *Bolero*. Working with a pair of composers, they whittled it down to four minutes and 28 seconds. Rather than reducing it further, Torvill and Dean challenged the rules by choreographing their sequence to start kneeling, with their skates not on the ice for the first 18 seconds of the music.

Shown live on television, over 20 million people in the UK watched with bated breath, since it was the only medal Great Britain would likely win in the entire winter Olympics.

Scoring perfect 6.0 ratings from the majority of the judges, Torvill and Dean achieved the highest score of any figure skaters for a single program in Olympic history – a record that stands to this day.

After, they built a career in ice-dance shows that culminated in the hit television series in 2006 called *Dancing on Ice*. It featured many two-person teams competing against one another. Each team was composed of an individual celebrity with little or no skating experience doing ice dancing with an ice skating professional. Torvill and Dean and the celebrity's professional ice skating partner taught the celebrities how to ice dance. The series has been franchised to over 18 countries, and in the UK has run for over 15 seasons.

The Williams Sisters

As for the other sports pairing, Venus and Serena Williams are two of the greatest women's tennis players in the history of the game. As a team, they are undoubtedly the greatest pair of tennis players in history.

Playing together as a doubles team, they won all 14 of the grand slam doubles finals they were in, with the first one being in 1999. They were each so individually much better than all other professional women tennis players at their peak that they reached the *same* grand slam final nine times. Serena won seven of them, but who won was secondary to how these sisters and teammates, and to how they transformed the sport of women's tennis and the race barriers they broke down along the way.

In the previously highly white, elitist world of professional tennis, together the Williams Sisters proved that Black women could win grand slam finals, Olympic medals in singles and doubles, and achieve Number 1 rankings in the world.

In doing so, their popularity, skill, competitiveness and tennis prowess brought women's tennis to prime-time television and increased the pay for women in grand slam tournaments. In 2007, for the first time ever, Wimbledon paid the women's winner, Venus Williams, the exact same amount as Roger Federer, the men's winner.

Their lives and their story, including that of how their mother and father supported and nurtured their training and careers were featured in the Hollywood film, *King Richard*. Like all the films made about the very small teams featured in this book, *King Richard* took some liberties in the story, but all the core elements of the film accurately depict Venus and Serena's rise from being underprivileged, discriminated against schoolgirls, to the very pinnacle of the tennis game. Together as a team they maintained that dominance of the sport for many years.

The most important team

There is one team we have not yet mentioned. It is the single most important team in literature, Hollywood, television and popular culture.

This team is fictional and is Arthur Conan Doyle's 1887 creation of the crime-solving duo, Sherlock Holmes and Dr. Watson. Conan Doyle wrote 56 short stories and four novels featuring this pair. Since then, more stories have been written by other authors and screenwriters using these two characters, and the original stories have been interpreted for screen and stage in a multitude of ways.

According to IMDb, there have been 52 different Holmes and Watson movies and television series made, with the earliest made in 1900.

Sherlock Holmes is a detective with an uncanny ability to detect clues, assess situations and people, and draw conclusions using logic that any ordinary detective (fictional or real) might miss. Holmes lives at 221b Baker Street – a fictional house number on a real London street.

The stories, as Doyle wrote them, involved Holmes' friend and assistant, Dr. Watson narrating most of them. Dr. Watson was more than a sidekick the way Robin is to Batman. Watson is able to spot and reveal clues that Holmes can work with. Much like the relationship between Kirk and Spock mentioned above, Watson brings human skills and emotion to the aloof and reserved Holmes who oftentimes acts in a mechanical, purely scientific, non-emotional manner.

The partnership works to catch thieves and criminal masterminds because, despite Holmes and Watson not being equals, they bring complementary skills to every endeavor, they value each other's company, insights and friendship, and most importantly, they share common goals.

These are important facets of any relationship, in our ordinary lives as well as in popular entertainment – the capacity to work well together, in spite of or perhaps because of, our differences, and leaven it all with a light-hearted touch.

★

Part 3 deals with three very small teams that emerged in risky and dangerous situations. They faced considerable challenges and in doing so, left an unforgettable legacy of leadership, teamwork and achievement.

If you ever think you are in a tight spot or a challenge looms ahead, these next three teams can serve as your inspiration.

Part 3

Dangerous Situations

★

Hillary and Tenzing on the South-East ridge about to leave for the South Col to establish Camp IX below the South Summit on Everest. (*Photographer: Alfred Gregory, May 28, 1953.*)
(*© Royal Geographical Society [with IBG]*)

Chapter 29
Mt. Everest

Hillary and Tenzing

"There is something about building up a comradeship - that I still
believe is the greatest of all feats - and sharing in the dangers
with your company of peers. It's the intense effort, the giving of
everything you've got. It's really a very pleasant sensation."

—*Edmund Hillary*

When the British mountaineer George Mallory was asked in 1924 why he was interested in climbing Mt. Everest, his reply was simply: "Because it's there."

However, it is much more than "there."

Mount Everest is situated in a region as challenged by politics as it is by geography and weather, and through the years all those factors have influenced climbers' ability to summit it. The Himalayan Mountain range in which Everest sits contains over one hundred imposing, towering and difficult-to-climb mountains, all of which have peaks over 24,000 feet (7,315 m) above sea level.

In a world where people had been attempting to reach the North Pole for centuries, and the South Pole ever since a human first set foot upon Antarctica in 1895, Mount Everest – the tallest mountain in the world – was naturally the next great geographical prize. Given that people have lived in its shadow almost since the dawn of humanity, it may seem surprising that the first attempts to climb it did not occur until the 1920's.

A team endeavor

Before we get into why that was the case, as well as the challenges of climbing such a forbidding and treacherous mountain, it is worth discussing mountaineering as a team endeavor.

Nowadays, solo mountain climbers give us spectacular achievements like those seen in the popular movie *Free Solo*, where Alex Honnold climbed the exposed rock face of El Capitan in California in 2017, alone and without fixed ropes. But in the early days of Himalayan mountaineering, tackling the biggest and most daunting climbs were team endeavors.

In the case of Everest, from the 1920s to 1950s, a two-person summit team was put into place near the summit via the efforts of a much larger team. Dozens or even hundreds of staff made up of other western climbers and local Sherpas supported each expedition. It took the ambition and preparation of the pair who made it to the top, to successfully utilize the resources of the broad pyramid of support workers below.

Our primary focus in this chapter is the team of Edmund Hillary and Tenzing Norgay who, together, were the first to reach the top of Everest in 1953. They were a two-man team within the larger British Mount Everest Expedition team, although neither was actually British. Hillary was a New Zealander; Tenzing was a Nepalese Sherpa.

The story of how they became a team within a team, and a look at the entire expedition's challenges and successes, reveals insights into how very small teams can perform within a bigger framework.

Why was climbing Mount Everest so hard?

According to the climbers of the era, Mount Everest did not possess the most challenging routes, nor was it the most challenging mountain. There were more difficult mountains and much more difficult climbs in the world, but Everest had specific challenges that had long thwarted small teams of great climbers.

The first difficulty was in gaining access.

Everest sits along the border of Nepal and Tibet, and for many years both countries were closed to foreigners. Even though the British had first sighted the mountain from India in the 1850s, and even calculated its height – declaring it the tallest on Earth and naming it after the British Surveyor General Sir George Everest – they did not have permission from either country to climb it. That was until the 1920s, when Tibet finally opened to the British.

Although there were excellent mountaineers of all nationalities, Mount Everest seemed reserved, and even destined, to be conquered by British climbers.

A few years after the end of WW2, the Tibetan government revoked their openness to foreigners. In 1949, neighboring Nepal opened their borders to select mountaineers from a variety of countries, to permit climbs from their side.

The British, Swiss, French and other nationalities formed teams to climb from the Nepalese side. In these early decades of climbing on Everest, a team was not allowed to choose which side, Tibet or Nepal, to begin their climb from. That was dictated by politics and governments at that time.

The next big challenge was getting the mountaineers and their needed supplies to the foot of the mountain.

Any plan for climbing required first establishing a lowest camp at the base of the mountain and then moving supplies along, setting up smaller camps along the way further up the mountain. Even today, there are no paved roads into or out of the base of the mountain. The track was even rougher in the early years. This meant all supplies – tents, sleeping bags, food, medical supplies, oxygen equipment, ropes and everything else needed – for a summit attempt would have to be carried on foot to even the lowest of base camps.

Determining how to assemble enough supplies and equipment, and choosing the team of western climbers and the Sherpas and other support staff all needed to be solved at the outset. The more people in the support team, the more food and supplies would be needed.

A mathematical puzzle

The leaders and planners needed to find the exact balance between people and supplies, and decide how high up the mountain each person or object would go, bearing in mind that people cannot stay at high elevations for extended periods of time. There had to be ongoing support teams ascending to establish higher level camps, and then descending later the same day, or soon after that, to recover at the lower altitudes of one of the other camps.

One of the many challenges in planning the early expeditions was where to site the base camp and the intermediate and upper-level camps. These are a crucial decision on any mountain but especially on Everest. Why? Because a major portion of the climb was above the "death zone", which is reached at 26,000 feet (7,900 m) above sea level. Once above that level and into the death zone, a person starts to deteriorate physically and mentally, and sometimes quite rapidly.

Also a considering factor was how high on the mountain would the uppermost camp have to be situated, since it was from there that the difficult single-day-there-and-back summit attempt would be launched. The higher up the mountain that the final camp was placed, the less distance the team needed to climb on summit day. But a camp at very high altitude in the oxygen-depleted air creates other challenges, as you will see later in this chapter.

Another key decision was the use of oxygen equipment.

Even in the 1920s, some expeditions brought along oxygen tanks for use in the rarefied air above 19,000 feet (5,790 m). There were two types of systems: closed and open. The type chosen might not only determine success or failure, but it could be the difference between life and death.

Who is on your team?

Every expedition leader has to ask these questions: who is going to be on your team? Who are your strongest, fittest climbers, the ones who are most surely to make the summit? Which of your mountaineers do their best work as a team?

How do you manage the larger team of Nepalese Sherpas and

other support staff who do not speak your language? Who among them will serve as your sirdar (head Sherpa)? How well will your team of highly motivated, and often individualistic, western mountaineers work with the Sherpas?

The expedition leader had to accept that even he may not be in the final two-man summit team. (In these early years all expedition leaders were men – Junko Tabei was the first woman to summit Everest with her ascent in 1975.) Any climber or expedition leader at any point may suffer from altitude sickness and other ailments that could force them to descend to the safety of lower altitudes. The strains and risks to the human body at death-zone altitudes are severe enough to bring on a heart attack, or pulmonary or cerebral edema without warning (a build-up of fluids in the climber's lungs or brain). To make matters worse, all these medical emergencies could happen at the same time to the same person.

These dangers meant the expedition leader had to be flexible in making decisions.

Selecting top climbers and summiters too early could result in the team not being fit enough on summit day. Selecting them too late could cause dissension in the ranks as people vie for position. Every decision posed risks to both the individuals and the expedition.

The cold and the altitude can create errors in judgment, tricking climbers into thinking they are stronger than they are, or that the summit is nearer than it is. If a climber keeps ascending at dusk when they should be descending the mountain, it can lead to serious injury or death. Descending in the dark is a recipe for disaster.

There were very real physical dangers at every turn – avalanches, hidden crevasses, near-vertical walls of solid white ice, frostbite, hypothermia, snow blindness, and these could occur all at once.

With so many challenges, it is a wonder that any of those early mountaineers managed to climb high on the mountain at all. That they achieved ascents above the 26,000 feet (7,900 m) death-zone level is a testament to their perseverance, hard work, climbing skill, and willingness to work together.

As you will see, many teams attempted it and made great progress, but only one highly remarkable pair of climbers got to the summit first.

Were Hillary and Tenzing the first?

This mystery is described at the end of Chapter 33, and it will be left for you to decide.

Chapter 30
Mt. Everest

The Most Handsome Man

In 1921, after Tibet opened access to their side of the mountain, an expedition, aptly named the First British Everest Reconnaissance Expedition, was launched. The team approached the challenge of climbing Everest in a systematic manner, starting with a smaller, less ambitious mission to understand the scale of the problem before attempting the summit.

The goal was to photograph and survey the mountain while seeking viable routes up from the north side and pioneering an overland route to the base of Everest. The latter would solve one of the first dilemmas: how to reach the mountain.

There were nine mountaineers accompanied by over 100 Sherpas and other support staff, but only eight westerners returned; one had a heart attack on the route leading to the mountain. He was the first but not the last of the casualties of the early expeditions.

Among the nine British climbers was a mountaineer of great talent, George Mallory. There are not many photographs of Mallory. Apparently none do him justice; he had been described by both women and men of the era as one the most handsome men anyone had ever seen. He was also an exceptionally skilled and brave mountaineer.

Mallory, working in a team with two other climbers, made the first-ever major ascent on Everest, reaching a height of 23,000 feet (7,000 m). There they discovered and named the North Col – a steep-sided glacier pass leading to the East Rongbuk Glacier – and determined that this was a potential route to the summit.

A new expedition was assembled in 1922 and named the Second British Everest Expedition. (You'll start to see a pattern emerge for British expedition names.) It included Mallory as well as other experienced mountaineers and John Noel, the expedition's photographer. They made considerable progress on the mountain using small teams of climbers.

The first team of four climbers reached 27,000 feet (8,230 m) before retreating. A few days later, a second team, aided by oxygen tanks, attempted the climb, achieving a high-altitude mark of 27,300 feet (8,320 m). After regrouping lower down the mountain, a few weeks later Mallory was leading a summit attempt when disaster struck.

In an avalanche, seven Sherpas died.

Everest may not have been the most difficult mountain to climb, but it was already claiming many lives, and this was only the beginning.

The Third British Everest Expedition

In 1924 the Third British Everest Expedition's two-man team of Edward Norton and Theodore Somervell reached 28,000 feet (8,500 m). Climbing without oxygen hampered their summit attempt, so Somervell turned back. Norton carried on alone, reaching a record high point of over 28,100 feet (8,560 m) before retreating. This was an extraordinary achievement.

To put it another way, Edward Norton in 1924, climbing the last portion alone and without oxygen, reached to within 1,000 feet (304 m) of the summit.

Two days later, Mallory teamed up with the young and exceptionally fit, but inexperienced climber Sandy Irvine who had previously rowed for Oxford University. Mallory and Irvine climbed using oxygen.

They both died on the mountain.

Mallory's body was found in 1999. Irvine's was never found.

To this day, the mystery about why, where, and how Mallory and Irvine fell on Everest is as great as the disappearance of Amelia Earhart and Fred Noonan on their round-the-world flight, thirteen years later.

More attempts, more failures

More British attempts were made in the 1930s. None reached higher on the mountain than Norton had achieved in 1924 without oxygen.

On the 1930s expeditions, important names and powerful small teams relevant to Hillary and Tenzing's 1953 summit success began to emerge. On the 1933 Fourth British Everest Expedition, the two-man team of Eric Shipton and Frank Smythe camped in the death zone successfully without oxygen, but their summit attempt failed.

Unrelated to that British expedition, 1933 was the year the first flights were made over Everest, and Orville Wright, who lived until 1948, would certainly have known about them.

The 1935 Fifth British Everest Expedition included British climber Bill Tilman and – in the first of his many expeditions – Sherpa Tenzing Norgay. That attempt also failed, as did the 1936 Sixth and the 1938 Seventh British Everest Expeditions. On that last expedition of the decade and the last British attempt from the Tibet side, Shipton and Smythe tried for the summit again, and failed.

A disturbing pattern was emerging.

Very small teams of highly accomplished British mountaineers were trying and failing, with oxygen or without. During the 1930s, however, Tenzing was gaining climbing experience that would prove invaluable later on. He was a Sherpa on the 1935, 1936 and 1938 expeditions.

In the early 1940s, with World War II raging, no serious attempts were made. Some less formal attempts were tried after the war, including a 1947 one which Tenzing participated in. And then Tibet closed its borders.

Entering the 1950s, Norton's 1924 Tibetan-side, non-oxygen-supported, highest altitude record stood uncontested.

New decade, new side of the mountain

In 1950, Bill Tilman and others surveyed the approach from the Nepalese side, looking for a new way to start.

Above the base of the mountain from that side looms the Khumbu Icefall. At 18,000 feet (5,490 m) above sea level, the Icefall is littered with enormous, shifting, unstable blocks of snow and ice. Pieces can be as large as a house, and these can be separated by chasms, cracks and crevasses, some reaching over 50 feet (15 m) wide and 100 feet (30.5 m) deep. Bridges of blown snow can form covering some of the gaps over, creating an unseen hazard.

A wrong step in the Khumbu Icefall region could mean serious injury or even death. It was treacherous and unforgiving. Stepping onto a snow bridge over a hidden crevasse could mean the mountaineer falling through as the snow bridge collapsed. Avalanches were also a constant peril.

Tilman initially thought the Icefall to be an impenetrable route for establishing supply lines to locations where upper-level camps would need to be erected. There appeared to be no way to make a summit attempt on Everest from the Nepalese side without the entire team and all the supplies traversing the Icefall. What was a viable route through the Icefall one day could be completely impassable the next, as a result of shifts in the unstable ice blocks.

Just like their counterparts did in 1921 on the Tibetan side, now thirty years later, the British set up a reconnaissance expedition to assess routes to the summit from the Nepalese side. Their first aim was to determine if Tilman was correct in his assessment of the Icefall.

They knew every attempt had to be precise because the Nepalese government announced they would only issue one Everest climbing permit per year.

The 1951 British Reconnaissance expedition

This 1951 British Reconnaissance expedition, led by Eric Shipton, included Tom Bourdillon, Ed Hillary, Michael Ward, and others who will feature prominently later in this chapter.

Shipton and his men trekked across the Khumbu Icefall avoiding the massive crevasses, almost reaching the relative safety of a glacier-formed valley that lay above it. This area had first been seen and named by Mallory on the 1921 reconnaissance expedition[13]. Mallory called it the Western Cwm (pronounced koom) after the Welsh

---

[13] The Western Cwm was visible to Mallory, even though he was climbing from the Tibetan side.

word for valley.

At various points, they divided into smaller teams consisting of two to nine men, to try different routes through the Icefall. Each route had its own challenges and obstacles. One area that had to be passed through was quickly dubbed the Atom Bomb Area because the vast array of large ice chunks resembled World War II bomb site debris. One climber, Earle Riddiford, slipped, saved only by the rope he was attached to. Later, two others narrowly avoided getting struck by tumbling ice and rocks. Weeks went by with only limited success.

Shipton thought they could make one final attempt before the climbing season came to a halt. He reasoned that a larger team could get through the Icefall and reach the Western Cwm. A team of nine men, including Shipton, Hillary, Bourdillon and Ward, two other western climbers, and three sherpas set out. They made good progress through the Icefall and the Atom Bomb Area only to be stopped by an enormous crevasse, seemingly impossible to cross. Knowing this was their last attempt before returning to Britain, they assessed the scene ahead of them.

What they saw thrilled them. They were staring at a possible route to the summit of Everest from the Nepal side.

It wasn't direct, but it could work.

It led from the Icefall to the Western Cwm, up to the South Col where at that point the climbers would be temporarily ascending on Lhotse, the fourth tallest mountain in the world. Once high up on Lhotse they could traverse across the South Col to place them high up onto the face of Everest, and then onwards to the summit at over 29,000 feet (8,839 m) above sea level.

It would not be easy but it was achievable, if only there was a way for a team to get across the wide crevasse stopping them at this very moment.

Competition: the Swiss versus the British, 1952

The Nepalese government gave the single 1952 climbing permit to the Swiss, who had been trying to get a climbing permit since the 1920s. There was talk of a joint Swiss-British attempt, but who would actually lead the expedition was hotly debated. The British proposed that the Swiss lead the joint expedition on the trek to Base Camp, with the Everest-experienced British mountaineer, Eric Shipton, leading the expedition to the summit.

Every team we studied for this book showed that successful teams need cohesion, clear leadership, and well-defined agreed goals. In 1952, a Swiss-British joint expedition would have none of these.

When the Swiss rejected the British proposal, it was decided that while the Swiss were at Mount Everest, the British, led by Shipton, would try to climb Cho Oyu, the sixth tallest mountain in the world, and located a mere 12 miles (19 km) from Everest.

At Cho Oyu, the British would continue to perfect their teamwork skills, and test out supplies, rations, tents, oxygen systems, clothing, boots, and other equipment. They would be primed for success on Everest the following year, if the Swiss were unsuccessful in 1952.

There was another win in it for the British. If the Swiss failed to reach the Everest summit, and the British succeeded reaching the Cho Oyu summit, the British would own the record for the tallest mountain ever climbed.

The win for the Swiss, if they succeeded, would be having climbed the tallest mountain in the world on their first attempt. By 1952, the British had eight attempts and none were successful.

The Swiss team had experienced mountaineers including an exceptionally strong climber named Raymond Lambert. Tenzing Norgay, with his extensive Everest experience, was selected as sirdar, the title for the man who would oversee the Sherpas.

At the top of the Icefall, they reached the same fear-inspiring, wide crevasse that had stopped the British the year before. The youngest of the Swiss climbers, Jean-Jacques Asper, wanted to attempt it immediately. He would be lowered into the crevasse on a rope to a point where it narrowed, and then try to swing across the gap. As he did, he would ram his ice axe to snare the other crevasse face, and then climb to the top of the crevasse. Once at the top he could use the rope to pull over the makings of a rope bridge, build it, and then get the other men and all their gear safely to his side.

It was risky, brave, and possibly even foolhardy.

The first try failed and Asper was slightly injured, but he was game to keep at it. They rigged a better rope system the next day and lowered Asper farther down the crevasse. This time his leap and swing across was successful.

Sometimes for a team to succeed, it takes the bravery of one individual team member. In this case, it was Asper.

From there they were the first to climb the route the British had seen in 1951 – the Icefall to the Western Cwm to the Lhotse face, across the South Col, and onto Everest.

The only common language

The 1952 summit attempt came from an unlikely very small team – the Swiss Raymond Lambert and the Sherpa sirdar Tenzing Norgay – the two strongest climbers, who partnered well together even though neither understood the language of the other.

Lambert and Tenzing were a formidable climbing duo. They shared a love of the mountain, and a camaraderie-based mutual trust and admiration for each other. After a sleepless night, high up on Everest, they then ascended to a new all-time high 28,210 feet (5.3 miles or 8.6 km above sea level) before having to turn back. Upon their return, Tenzing was invited to be an honorary member of the Swiss Alpine Club.

Meanwhile, the British were failing on Cho Oyu, though they were learning many important lessons for their upcoming 1953 attempt at Everest. They were finding that Eric Shipton, even with his many years of Everest experience, was too individualistic to lead the large team effort that the British mountaineering community felt would be needed to summit Everest.

The push to reach the top of the highest mountain in the world was becoming a race with nationalistic overtones. The Swiss had failed in 1952; if the British failed in 1953, the French would be next to try. They had already obtained permission to attempt it in 1954.

It was into this competitive situation that the British formed a 1953 Everest expedition.

Chapter 31
Mt. Everest

Finding The Right Leader

Eric Shipton would have been the natural choice for leader of the 1953 British attempt, but the Cho Oyu expedition raised too many questions about his leadership. It was now understood that reaching the top would require managing a very large team of climbers, Sherpas and other staff, as well as an enormous quantity of supplies. An operation this large would need to be run with military precision. Out of the many possibilities for the role emerged Colonel John Hunt.

Hunt had many redeeming qualities and one serious drawback. He was a mountain climber and skier, but when he was originally invited to apply to be part of the 1936 British Everest attempt, it was discovered he had a heart murmur. He never even made it to the short list of contenders to join that expedition. At death zone altitudes, even a minor heart condition could become serious. But that didn't stop Hunt from climbing many mountains, including others in the Himalayas. Despite his heart murmur, he was an established mountaineer.

Putting his military campaign expertise to work, Hunt focused on building a solid team with a single purpose: to be the first to climb to the top of the highest mountain in the world. There were to be no distractions involving science, mapping or anything else.

Assembling the team

Hunt's initial aim was to find ten or eleven men who had not only the right experience but also the ability to work collaboratively with others, who had an innate love of mountaineering, and who were British. He later expanded this final attribute to encompass the entire British Commonwealth of nations, so he could include two strong climbers from New Zealand who partnered well together: Ed Hillary and George Lowe. Hunt also believed that all team members should be 25 years or older, but also relaxed that rule for two of the team.

In alphabetical order, Hunt's team consisted of George Band (one of the under 25-year-olds), Tom Bourdillon (oxygen tank expert), Dr. Charles Evans, Alfred Gregory, Ed Hillary, George Lowe, Wilfred Noyce, Dr. Michael Ward (expedition doctor), Mike Westmacott, and Charles Wylie.

In addition to the core team of western climbers Hunt invited a physiologist, Griffith Pugh, to advise the expedition on the challenge of acclimatizing in high altitudes. Here again Hunt was relaxing one his initial rules. Pugh was also going to be doing research on the climbers' fitness, adding a scientific goal to the expedition. Tom Stobbart was the expedition photographer. Tenzing Norgay was appointed as sirdar, head of the Sherpas.

Hunt was a military man by training and career, but he was proving that he could be flexible in his thinking. This could become strength or weakness on a mountain as unforgiving as Everest. Hunt understood the value of expedition publicity, both before and after the hoped-for successful ascent. A successful summit would a front-page story seen around the world.

He arranged for the rights of the story to be sold to a British newspaper, *The Times,* which sent reporters Arthur Hutchinson and James Morris to accompany the expedition. Hutchinson was to be based at Kathmandu relaying reports to *The Times* London headquarters; Morris was to be embedded with the climbers as far as he could go. Another reporter, Ralph Izzard, from the rival paper the *Daily Mail*, was also vying for a scoop. None of these reporters were official team members but they became part of the story.

The scale model

Selecting the route and understanding the terrain of Everest and the surrounding mountains were vital to a successful mission. A great team and a poor route would not lead to a successful and safe climb. To aid in the process, a scale model of Mt. Everest and its surroundings was built to exacting standards.

All the accumulated knowledge from 1920s expeditions, the 1933 flights over Everest, and the 1951 British Reconnaissance expedition was compiled to make the model. A scale of 10 inches to 1 mile (25cm to 1.6 km) was used. Eric Shipton, though not selected to lead the 1953 expedition, provided valuable input. The model, which measured 6.5 feet (2 m) in width and length, was built with plaster on a timber base and housed at the Royal Geographical Society, where it is still on view today.

John Hunt (who had actually never climbed on Everest even at lower altitudes) and the team used the model to visualize the route they would be taking – the one seen and thought possible by Shipton, Bourdillon, Hillary and Ward in 1951, and attempted by the Swiss in 1952. It led from the Khumbu Icefall to the Western Cwm, up the South Col on Lhotse, a traverse across the South Col over to Everest, and then upwards to the Summit.

It seemed daunting yet doable.

The question was who among the ten-man core team would be the climbing pair to make the first summit attempt?

To a strong climber like Ed Hillary, it seemed unlikely that he and George Lowe, New Zealander climbing partners, would be the first ascent team on a British expedition, no matter how strong and experienced they were.

The start

Assembling the team, route and supplies on paper was one thing. Amassing all personnel, equipment, oxygen, food, tents and other supplies in Nepal was more of a challenge. They needed almost 10,000 pounds (4,500 kg) of equipment brought to the base of the mountain and up into the higher elevations to create camps. They would be using a broader team of Sherpas, porters and local villagers in the early stage, and then mostly Sherpas for the higher altitudes.

They all arrived in Kathmandu in mid-March 1953. Kathmandu sits at 4,300 feet (1,320 m) above sea level. While lower in elevation than some US cities, it is an elevation that for some still requires some acclimatization. The month-long journey from Kathmandu to the base of Everest involves a rise in elevation by over 10,000 feet (3,050 m).

The mountain journey began in earnest on April 9, with Hunt designating Hillary as the leader to take a team of five Sherpas, 39 villagers serving as porters, and the team of George Band, George Lowe, Griffith Pugh, Tom Stobart, and Mike Westmacott to the location on the Khumbu Glacier that would become their Everest Base Camp at 17,500 feet (5,300 m).[14]

As with everything on this expedition, there were challenges. A snowfall one night and dazzling sunshine the next day meant everyone suffered from snow blindness, a terribly painful affliction only preventable through the use of snow goggles. The British team and Sherpas had snow goggles; the villagers did not. Stobart's quick ability to improvise led to the creation of makeshift goggles for all who needed them.

Once at Base Camp the villagers departed, having served their duty. By April 20, the entire British team arrived there, many of them suffering various ailments like coughs, sore throats and diarrhea.

In a surprising twist, Ralph Izzard, the *Daily Mail* correspondent who shouldn't have even been on the mountain, hired a few Sherpas and made the journey to Base Camp.

He was not welcomed there. The expedition's story was already committed to *The Times*.

The mountaineers had to grudgingly give Izzard, an untrained climber with insufficient experience or gear, some credit for having achieved such an altitude. It was clear to some that while the battle

---

[14] The 1953 Everest Base Camp was over 1,500 feet (460 m) higher in altitude than Mont Blanc, the tallest mountain in the Alps. The Everest Base Camp was also at a higher elevation than the top of every mountain in North America, except Denali.

between the mountaineers and the mountain would persist over the coming months, a battle between Izzard and James Morris of *The Times* would persist as well, with Izzard hoping to nab the scoop of a successful climb to the top of the world.

Chapter 32
Mt. Everest

Hell Fire Alley

Illary and the others surveyed the Icefall, their first big obstacle. It looked far worse and more impassable than what they had seen on their 1951 British Reconnaissance Expedition.

Not only would they need to find a route through the enormous, shifting terrain of snow and ice blocks, but that route would need to stay stable and navigable for as long as the expedition would need it. It would be through this route that many climbers and Sherpas would bring three tons (2,720 kg) of supplies.

They took perverse pleasure in naming the worst sections of the Icefall. Mike's Horror, Hell Fire Alley, The Nutcracker and The Ghastly Crevasse provide a glimpse into their view of the Icefall. Using a combination of flags to mark the path, fixed ropes for handholds in tricky places, and aluminum ladders to cross crevasses, they created a passable route.

Crossing the great crevasse

The great crevasse that had stopped the British Everest Expedition in 1951, and was crossed at great risk by Jean-Jacques Asper on the 1952 Swiss Everest Expedition still lay ahead. The 1953 team took a different approach to cross it.

Three aluminum ladders, bolted end-to-end, were barely long enough to reach from one side of the crevasse to the other. Who would be brave enough to be the first to cross?

Hillary roped up, attached to Lowe, Tenzing and Wylie. If the ladders bent, swayed too much, or the joints snapped, Hillary would fall. Would the three men get pulled in as well, or would they be able to dig in, to save the tall New Zealander from further calamity?

All went well, and soon they were standing on the Western Cwm.

Every journey up and down the mountain involved crossing that crevasse. Getting supplies through the Icefall was a team effort involving Sherpas and western climbers. Each man had to make many trips. Teams of Sherpas were always accompanied through the Icefall by at least one of Hunt's men. Goods were brought to the edge of the great crevasse and stockpiled, waiting for the strongest, most agile and experienced men to carry them across the aluminum set of ladders.

Once across, the supplies could be carried up to what they called the Advance Camp, which was situated in the Western Cwm.

Near the crevasse and beyond, the team was delighted to discover rations had been left behind by the Swiss team the year before. These were far better than the British rations.

A team within a team

By late April, Advance Camp was well established and the task to move supplies higher on the mountain was underway.

The prize for every climber was to be in the two-man team climbing for glory – the summit – on the determined day. It was surely a large team effort to get to this point, but to every mountaineer, there was a special place in their heart to be the first to summit the tallest mountain in the world. It was clear an attempt could be made in May.

Hillary knew that a British expedition would start by trying to put a British two-man team on the summit. That ruled him out of being in the starting two. He reasoned that he could be in the second pairing.

It seemed unlikely that John Hunt would be one of the summiters given his consistent health issues. Several other men were ailing at times. Pugh and Stobart were not the core climbing team, so both seemed unlikely candidates. Lowe, also being a New Zealander,

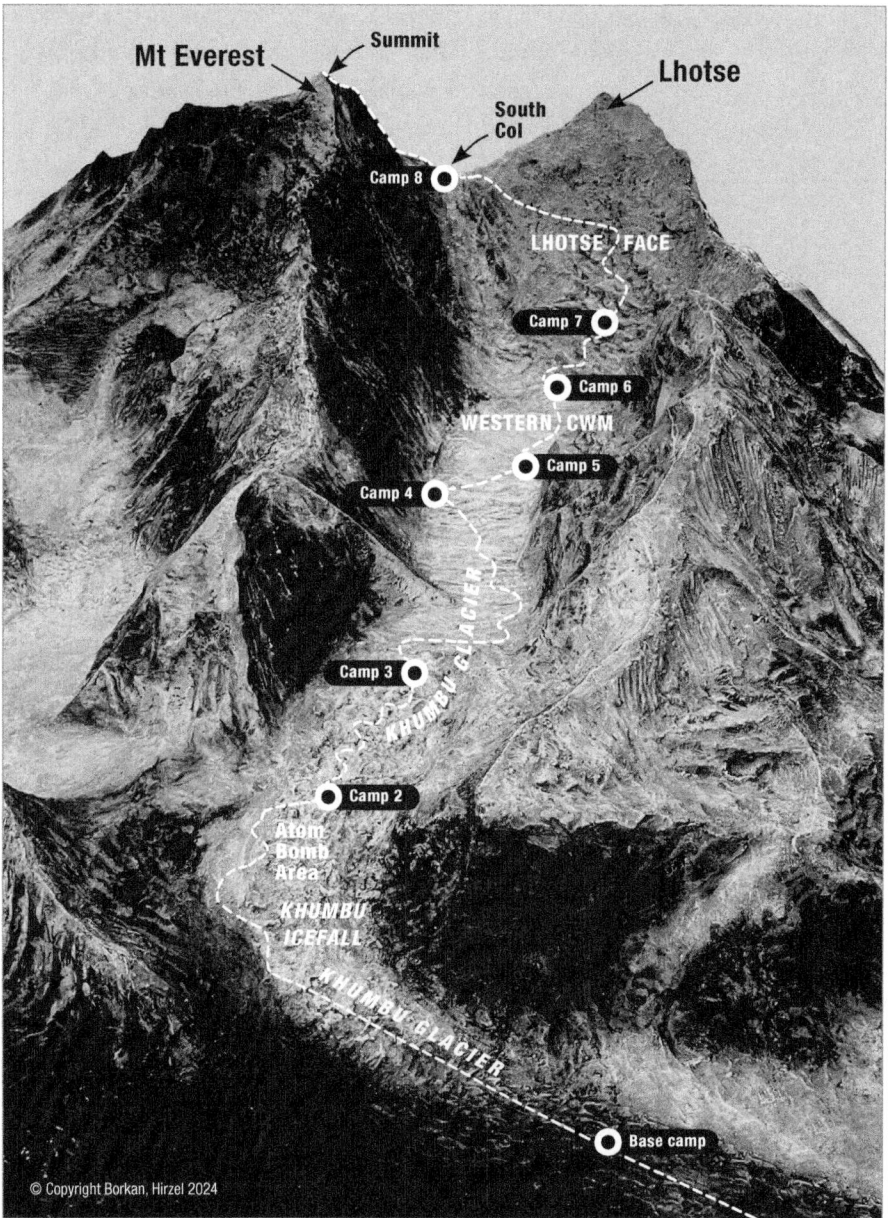

1953 route superimposed on a photograph of the scale model at the Royal Geographical Society. (*Photographer: Brad Borkan*)

could be a competitor for the second team, but not as a partner to Hillary. After all, how would a British expedition be received back home if the two men on the top were both New Zealanders?

Though Hillary had a working relationship with Tenzing, so did all the other men of the British team. Tenzing was enormously accomplished as a climber and had a serious determination to reach the top. He had much more affinity to the 1952 team of Swiss climbers and particularly Raymond Lambert, the Swiss climber he had teamed with to set the Everest high-altitude record.

On the morning of April 26, Hillary, Hunt, Tenzing and Wylie were back at Base Camp. They decided to head for the Advance Camp, which meant trekking and climbing through the length of the Icefall and ascending in elevation a considerable amount. Hillary and Tenzing set out strongly, leaving Hunt and Wylie to catch up.

Having reached Advance Camp, Hillary and Tenzing decided a bit later in the day to descend back to Base Camp. Such an upward and downward hike was far more climbing than was wise to attempt in a single day. The ascent was difficult, and the descent was brutal in other ways. At one point Hillary leapt across a crevasse and misjudged the distance. Falling in, he was only saved by Tenzing's quick reflexes steadying himself and the rope that joined them.

From that moment, an unbreakable bond was formed between them.

Lhotse

Lhotse is one of the mountains adjacent to Everest. The route planned in 1951 and now being executed was to climb the Western Cwm to arrive on Lhotse at a height of 22,000 feet (6,705 m), then to establish camps on Lhotse and climb to 26,000 feet (7,900 m) (a death-zone altitude) before crossing the South Col onto Everest.

At these altitudes oxygen would be needed, but the tanks were heavy and limited in numbers. They needed to be rationed and used carefully, to ensure enough for the summit attempts. Bourdillon was the expert on the systems. They had two types of systems: Closed and Open.

The Closed system involved the climber carrying two canisters with a total weight of 45 pounds (20.4 kg), one filled with oxygen that the climber inhaled. The other contained soda lime, which captured the carbon dioxide the climber exhaled, and then recycled it.

The advantage of the Closed System was it gave a strong flow of oxygen to the mountaineer or Sherpa. The user could climb faster, as he would feel like he was at a much lower altitude. That, in theory, would offset the greater weight of the two-tank system. One of the main downsides was that if the climber had to remove the mask, he would find himself less acclimatized to the altitude he was at. Bourdillon and Charles Evans used the Closed system on a trek from Advance Camp down to Base Camp and found it performed well. This was the system Bourdillon preferred.

The Open system combined oxygen with outside air. It required a single tank since no soda lime tank was needed. This was the system used by Mallory and Irvine in their 1924 attempt, and tested by Hillary and Tenzing on their Base Camp to Advance Camp climb. They completed that climb in five hours, which was a very good time.

The choice of systems played a key role in what happened next.

Chapter 33
Mt. Everest

Could They Do It?

Numerous challenges faced the climbers, some brought on by the weather, others resulting from a combination of bad luck and poor decision making.

Without enough oxygen tanks for everyone and the need to move one ton (1,000 kg) of supplies up Lhotse, Hunt decided the Sherpas would have to get by without oxygen. At these altitudes it is a hard enough climb, even with no pack at all. The Sherpas carried sizable packs, weighing up to 50 pounds (23 kg) each.

To ease the trek up Lhotse, Hunt appointed George Lowe and Sherpa Ang Nyima to cut all the steps in the ice needed to get up to 26,000 feet (7,900 m) on Lhotse in a week. This task would likely take Lowe out of the running to be a summiter. But like everyone, he understood that the support team was all important to the success of the expedition, even when the time allotted to the task was ridiculously short.

Setbacks included a snowstorm on Lhotse, which hampered the work. Two climbers – George Band and Mike Westmacott – became ill and weak. Add to that, the Icefall was shifting in ways that would compromise the return route. Halfway through the tedious, strenuous work of cutting steps, George Lowe became sick.

With the steps half done and the weather window for summit attempts rapidly closing, Hunt faced a difficult decision. He could either send a potential summit person to take Lowe's place digging steps or he could send Sherpas laden with heavy packs and equipment to traverse the steps that had been dug, and continue onwards up Lhotse and onto the South Col on the unfinished trail.

The decision was to do the latter and 14 Sherpas were sent in two groups of seven. They all climbed without any oxygen. This was a terrible and arduous task, but it was successful.

By May 22, there were 17 men on the South Col, including Hillary, Noyce and Tenzing.

The summit teams

John Hunt's summit plan was emerging. Despite the illnesses, he still had excellent acclimatized climbers to choose from.

Choosing the starting point for the summit attempt required Hunt to balance several factors. The two men selected for the summit team would be supported by two Sherpas. Together they would ascend to the agreed elevation and establish their final South Col camp. The team would have to overnight at death-zone altitude, requiring sleeping with an oxygen mask that would likely interfere with a good night's slumber. If the camp is too low, the climb to the summit is too long. If the camp is situated higher on the mountain, then sleeping is harder, and risks increase. With oxygen tanks being so heavy, they would have to carefully ration what was used for sleeping and climbing.

Hunt decided that Bourdillon and Evans would make the first attempt. Together with two Sherpas, they set up camp on the South Col at 25,890 feet (7,900 m) on May 24. This was the eighth camp the 1953 expedition had erected on the route.

Bourdillon and Evans would need to ascend another 3,000 feet (914 m) to achieve the summit, and then turn around and descend the same amount to their camp, all in one day. They would use the Closed oxygen system for this daunting climb.

If Bourdillon and Evans did not make it, then it would be Hillary and Tenzing's turn.

Hillary's plan to prove that he and Tenzing were a strong climbing duo had paid off.

They would use the Open oxygen system and they would camp higher on the mountain at 27,390 feet (8,350 m). Their camp would have to be higher than that of Bourdillon and Evans because the Open system did not have the range to support a longer climb to the summit and back.

On May 26, there was every reason to be optimistic about Bourdillon and Evans' attempt. They were an excellent team, proven climbers, and they knew more about the oxygen systems than anyone else. Success depended upon the oxygen system working well. Unfortunately, problems with their system delayed their 6am start to 8am.

As a team they climbed well, passing the 1952 camp set up by Raymond Lambert and Tenzing. Three hours into the climb, Evans recommended they change the oxygen and soda lime tanks for fresh ones, even though they were removing tanks that were not yet depleted. This may have been a good decision to change canisters before they reached more exposed and treacherous conditions. The new canisters would give them 5 hours of climbing time, so with an estimated 1.5 hours to reach the summit, they would have excess in the tanks by the time they returned to their camp after the summit.

But Evans' canister proved to be defective. Conditions on the mountain steadily worsened and the 1.5 hours to ascend turned into 3 hours to reach a point high on the South Col, within 300 feet (92 m) of the summit. The last 300 feet may not sound like much on a mountain over 29,000 feet (8,839 m) tall, but at that altitude, in those conditions and with limited oxygen left, Evans insisted they turn around. Bourdillon wanted to carry on.

It was possible for their team to split up, just as Norton had done in 1924, continuing on while Somervell descended. However, this was 1953 and theirs was a different climb with a different sentiment. Hunt had always insisted this was a team effort.

Bourdillon could have continued alone toward the summit, leaving Evans with a very risky solo return to the camp on a route best done with a belay. But Evans' oxygen system was giving trouble, and Bourdillon was the only one with the skills to repair it. If Bourdillon reached the summit, he would also have to face the dangerous trek back to camp alone.

In a decision as important and life affirming as Shackleton and Wild's on their return from near the South Pole (described in Chapter 38, when they also chose survival over achievement), Bourdillon and Evans as a team turned around.

It was a decision that plagued Bourdillon for many years to come.

Their turn

Despite slips and falls, Bourdillon and Evans made it safely back to the high camp. There, they met Hunt, Hillary and Tenzing. Bourdillon and Evans had proven they were one of the great mountaineering teams. They may not have reached the summit, but in a single day, at the highest altitudes in the world, they had ascended 2,700 feet (823 m) and descended the same amount. It was a monumental climbing achievement.

The next day, Bourdillon and Evans would continue their descent, and Hunt would stay with Hillary and Tenzing. They would be joined by Alfred Gregory, George Lowe, and Sherpa Ang Nyima to help Hillary and Tenzing prepare for their summit attempt.

With limited tent space, supplies and oxygen, a frank discussion was had with Hunt, who was not well. As expedition leader, he had spent far too long at high altitude. The time had come for him to descend the mountain and make room at this high camp for a fresher team member who could assist with preparations.

In one of the most courageous and gracious decisions on the 1953 British Everest Expedition, Hunt agreed to descend from the high camp. On their expeditions, Shackleton (South Pole attempt) and Peary (North Pole attempt) led from the front. On Everest, on the cusp of the greatest triumph ever, the leader Hunt put the team first.

Hunt's departing advice to Hillary and Tenzing was to put safety first, but try their "damnedest" to get to the summit.

The next morning, May 27, Gregory, Lowe and Nyima started cutting steps for the upward path. The plan was for Hillary and Tenzing to follow them and reach the planned altitude of 27,395 (8,350 m) or higher to set up the highest camp on the mountain. They'd overnight on the mountain and climb to the summit on the next day.

After helping set up the last camp even higher on the mountain at 27,640 feet (8,425 m), Gregory and Lowe descended.

After a restless night, Hillary and Tenzing set out on May 28 at 6:30am. They reached 28,000 feet (8,535 m) and found the partially full oxygen canisters that Bourdillon and Evans had discarded too early in their climb. This valuable find gave Hillary and Tenzing an additional hour of oxygen.

By 9am they reached what was called the South Summit and they kept climbing. Hillary was carefully monitoring the oxygen, at one point having to clear Tenzing's outlet valve of his mask. It was a treacherous and tiring climb.

They kept moving upward, reaching an even more challenging obstacle – a rocky, icy outcrop that later had been dubbed the Hillary Step[15]. Every time they thought the crest before them would reveal itself to be the summit, another crest came into view. At last, at 11:30am on May 29, 1953, there were no more higher crests visible.

They had reached the summit.

After an exuberant embrace at the top of the world, Tenzing spread out a line of prayer flags while Hillary photographed him, the flags, and the surrounding area.

There are no photos of Hillary at the summit. Tenzing didn't have a camera, and actually had never taken a photo in his life.

They stayed on the summit for 18 minutes, time enough to place some personal possessions in the snow and look for any signs that Mallory and Irvine had reached it in 1924. They saw none. Even if Mallory and Irvine had reached it, any memento would have long since blown away.

The descent

They began their descent from the summit just before noontime,

---

[15] The rock formation of the Hillary Step changed after 2015 Nepal earthquake. The Hillary Step no longer exists in the way that Hillary and Tenzing experienced it.

reaching the South Summit at 1pm, and the high camp at 2pm, where they stopped to swap to the oxygen tanks left by Bourdillon and Evans.

By 4pm they reached the South Col and met up with Lowe, Noyce and Sherpa Pasang Phutar, and the next day began their descent to Lhotse and then on down to Advance Base.

Now a new race was on. They knew Queen Elizabeth's Coronation was scheduled for June 2. What a triumph it would be to break the news of the successful summit by the team of the 1953 British Everest Expedition and the very small team of Hillary and Tenzing standing on top of the highest peak in the world on the day of Coronation.

With the competition between *The Times* and the *Daily Mail*, and wanting to control the story, the expedition had worked out codes for relaying information.

"Snow conditions bad"

The code had been given in advance to the British Ambassador for Nepal. The key words that would be relayed by James Morris, *The Times* correspondent, would either be:

Snow conditions bad = success

or *Wind still troublesome* = failure

The full telegram sent read:

> *Snow conditions bad stop advanced base abandoned yesterday stop awaiting improvement. All well.*

The code reached London in the early hours of Coronation morning, on June 2, 1953.

But were they first?

Were Hillary and Tenzing the first on the summit, or did Mallory and Irvine reach it in 1924? Mallory and Irvine had been spotted at a very high location on the mountain by John Noel, one of their climbing partners, who was watching for them through a telescope.

Mallory was approaching the summit from the Tibet side, which involved climbing features that were different to those Hillary and Tenzing faced. Mallory's companions said he was carrying a photograph of his wife that he would leave at the summit, but when his body was discovered in 1999, no photograph was found. There was also speculation that had he reached the summit, he would have carried down some summit rocks in his pocket. This was something he had talked about. No summit rocks were found on his person.

The real proof would have been in the camera that Mallory or Irvine carried, but so far Irvine's body has not been discovered and no camera has been found. It is still unclear how they died. Did Mallory slip, pulling Irvine down the mountain with him?

Did they make it? No one knows. In a book about the 1924 expedition, John Noel wrote that he believed they could have reached the summit.

The measure of a true team

Immediately upon returning to western civilization, the questions began. Did Hillary set foot on the summit before Tenzing or was it the other way around?

Hillary and Tenzing refused to be drawn into the discussion. It was a team effort. Neither could have done it without the other. And the two of them could not have done without Hunt's leadership, the entire team of western climbers, and the many brave Sherpas, porters and support staff.

But reporters wouldn't give up.

Hillary's more detailed answer explained that on May 29, 1953, they kept trekking to the top of a ridge, and each time they reached what seemed like the top, there was a higher ridge to climb. One of the climbers led then the other. It was immaterial who was leading when they reached the highest level. There could just as easily have been another higher ridge to ascend.

Many years later, Hillary confirmed that he had indeed been the first, but in no way wanted credit for that. It was only luck of the draw as to who was in the lead at that moment.

Another way to look at it might be this.

The British climbing community's ambition was to put together a large team that would get this very small team to the top of

Everest. It was a true team effort, every single step of the way, to the very summit of the tallest mountain in the world. Every team member played a valuable and essential role, but it took two unique and determined individuals, from two very different countries and backgrounds, to team together to achieve the near impossible – success on Everest.

★

From high up on mountains to the upper atmosphere of space, very small teams have persevered. But not everything always goes according to plan. When the worst happens, leadership and, teamwork coupled with camaraderie is needed, and even then there's a risk of overshooting the moon and never returning.

Night time, ground level view showing the Apollo 13 spacecraft and
Saturn V rocket stages. March 24, 1970. *(Photograph: NASA)*

Chapter 34
Outer Space

Apollo 13

"I have never lost an American in space, sure as hell aren't going to lose one now. This crew is coming home. You got to believe it. Your team must believe it. And we must make it happen."

—*Gene Kranz, Apollo 13 flight director in Houston*

There are few teams as highly specialized as those chosen for the early NASA missions.

"Space is the new frontier," declared President Kennedy in September 1962, invoking a pioneer spirit infused with urgency and destiny. "We choose to go to the moon," and he promised to land men there and bring them home again by the end of that decade. The urgency in this audacious goal was triggered by the Soviet Union's resounding success so far in the "Space Race."

NASA, the civilian National Aeronautics and Space Administration, had only been formed a few years before. It was endowed with a limited program to explore space that mostly focused on launching unmanned satellites.

By the time of Kennedy's speech, the Russians were already far ahead in the space race. They had already put into orbit the first Sputnik satellite in 1957, followed by Sputnik II carrying the dog

Laika. Shockingly there were no plans to return Laika to Earth.

And then in April 1961, the Russian cosmonaut Yuri Gagarin was the first human not only to go into space, but to orbit the Earth on a flight lasting 108 minutes. Fortunately, there were plans for his safe return.

So far, all the United States had to show for their space efforts were two failed satellite launches, followed by the successful launch of the unmanned Explorer 1 satellite, and then the first US manned sub-orbital flight. Unlike what the Russian Gagarin achieved, the first American sent into space, astronaut Alan Shepard in the tiny Mercury capsule, didn't complete even a single orbit of the Earth in his 15 minute flight.

In trying to fulfil Kennedy's promise, the United States had an enormous challenge to face, one that would take a vast a team effort to achieve. There would be brilliant successes and tragic failures along the way, but at the heart of them all were the very small teams who risked their lives on some of the most complex and dangerous missions ever conceived.

This is the story of one of those teams.

Little margin for error

The earliest launch vehicles available to NASA were the Atlas and Redstone rockets. These had been developed as instruments of war, intended to launch nuclear warheads toward far-away American enemies. To make good on Kennedy's promise, these would be repurposed for the peaceful activity of propelling manned space capsules into outer space.

The first of these was the one-man Mercury capsule intended for short-duration flights, but plans were well under way for longer orbital missions with the two-man Gemini, and later three-man Apollo capsules, already in motion. These capsules, the rockets that launched them and the guidance systems that controlled them were designed long before the advent of modern computers. At the time of the early NASA space missions, the computer they used was the IBM System/360 Model 75 mainframe. Today, a single, ordinary cell phone has millions of times more computing power than that IBM computer.[16]

The US space program began moving forward rapidly. There was a growing need for highly specialized teams of astronauts to go in these capsules, trained to work together as a unit to pilot their spaceships through ever-longer and ever more specialized missions. These space journeys would become increasingly complicated and dangerous. Every astronaut team would face instant death should any of the many life-sustaining components fail.

Among these were oxygen tanks to breathe, CO_2 scrubbers to clear away the accumulation of poisonous gas, batteries to operate the electrical systems and warm the command, radios to maintain contact with Mission Control in Houston, fuel for the engines, and food and water for the men. The loss of any one of these components would doom any space mission and the astronauts on board. It was clear to all involved how little margin of error existed.

The astronauts were selected from a highly specialized field – that of test pilots experienced in flying innovative prototypes of new military airplanes. These test pilots were chosen from an even more elite group – jet-fighter pilots with combat experience, who already knew how to operate a fast-moving vessel, making lightning-quick decisions under extreme duress in situations where one wrong move could mean getting shot down.

Who better experienced than these to become spaceship commander and crew?

But the Russians still had a head start, both in terms of capability and public awareness.

Valentina Tereshkova, a Soviet woman cosmonaut, was the first woman to go to space. On a solo mission in 1963, Tereshkova made 48 orbits of the Earth on a space flight lasting three days. It was the maiden voyage of the Vostok 6, and to this day, Tereshkova is the *only* woman to have ever had a solo space flight.

---

[16] The IBM System/360 Model 75 mainframe cost millions of dollars and was so large in size, it came with a manual explaining the room and air conditioning requirements to house the computer.

It would take the United States twenty years later to put a woman in space, with Sally Ride aboard the Space Shuttle Challenger in 1983, accompanied by four men. It was the Challenger's second flight, not its maiden voyage.

Building the team

The Mercury missions proved it was feasible to send men into space. The Gemini program added longer missions with ever-greater complexity, but these were at best Earth orbital missions, never straying far from our planet. The three-man Apollo missions, building on the successes of the Mercury and Gemini programs, would be the ones to leave the Earth's orbit, for a possible landing on the moon.

To the moon and back – a mission far more complex and dangerous than its predecessors. If any of the Apollo astronauts-in-training dismissed the risks before 1967, they sure knew them first-hand on January 27th of that year.

The three men of the Apollo 1 astronaut team, Gus Grissom, Edward White and Roger Chaffee burned to death when a flash fire occurred during a pre-flight test conducted on the launch pad. They died in their space suits while seated in the Command Module. The simulation was taking place three-and-a-half weeks before their scheduled lift off. Theirs was to have been the first of a succession of Apollo missions – low-Earth orbits to test out the functioning of the Apollo Command Module and Service Module, and a stepping stone to ever-longer and more complex flights.

All missions required the best of the best in personnel to prove what had been only theoretical up until now – reaching the moon. That was until the space flight of Apollo 8

In December 1968 the three-man Apollo 8 team, Frank Borman, James Lovell and William Anders, were the first astronauts to leave the gravitational field of the Earth, going farther into space than anyone had ever gone before. They were the first to see Earth as a big, blue marble floating in the vastness of space and the first to see the moon close-up from a low-lunar orbit a mere 60 miles

(96 km) above the surface. Part of their mission was to test out the many systems of the intricate Apollo spacecraft. Everything went according to plan.

The Apollo program was designed so that each man had his specialty. Each of the separate teams intended for the upcoming moon launches trained together as a three-man unit. They all would spend hundreds of hours on the ground learning, training and preparing for every hour they would eventually spend in space. The cockpit of the training simulator was an exact replica of the interior of the spaceship's three modules, with every switch and dial in the control panel, all seats and windows, and the food lockers and toilets in the cramped space in the exact spots they would be found in the actual vessel.

Three men for the mission

For Apollo 13 one of those three men, the commander of the mission Jim Lovell had first orbited the moon in Apollo 8. He had more personal experience than most of the other pilots in the Apollo program and was on his way in Apollo 13, for this third-ever lunar landing, to walk on the surface of the moon.

Lovell built his own rockets as a boy in 1945, before graduating from the United States Naval Academy to become a fighter pilot and test pilot. He married the girl back home, and was chosen for NASA's Gemini-Apollo program in 1962. Apollo 13 would be his fourth flight into space, twice in the Gemini capsule, and now twice to the moon in Apollo.

At thirty-six Fred Haise was the youngest man on the Apollo 13 team. He was also a rookie in the lunar program. Haise would accompany Lovell to the surface of the moon in the two-part Lunar Excursion Module (LEM). Once in low-lunar orbit the two men would move from the Command Module into the LEM and descend to the moon while the Command Module continued to circle the moon.

Upon landing, Lovell and Haise would exit the LEM, explore the surface, and collect samples. When finished they would climb back in and using only the ascent stage of the LEM, return into space to rendezvous with the Command Module.

Reunited, the two vessels functioning as one would leave their lunar orbit and complete the two-day journey back to Earth. Haise, a former Marine pilot, made it his business to know as much or more about the LEM than its designers and builders.

The third man on the team, Jack Swigert, was also a rookie in the lunar program, but his training had not been with Lovell and Haise. This former Air Force pilot was chosen at the last minute to replace Command Module pilot Ken Mattingly, who had been exposed to German measles. As a result, Mattingly was scrubbed from the mission only four days prior to the scheduled launch on April 11, 1970.

Swigert would be the one to pilot the Command Module while in lunar orbit, while Lovell and Haise descended to the surface in the LEM to become the seventh and eighth men to ever set foot there.

Training for any emergency

The thousands of hours spent by the astronauts-in-training in solving the myriad of problems Mission Control could throw at them were intended to help them learn to function as a coherent unit rather than as three men thrown together as a result of their career choices. During training, in addition to rehearsing the many real maneuvers they would use during the flight, they were thrown into various "emergencies" crafted by the Ground Control team to perfect their responses to unexpected situations. These simulations were as close as possible to the actual flight and they demanded their in-the-moment attention, and flawless cooperation and execution.

In the flight simulator, the men learned to read the nuances of each other's voices, that essential human capacity to instinctively understand not only the meaning of the words spoken but how urgent their response needed to be. They also learned how to deal with the frustrations felt when the demands of the ground crew seemed to ignore the competence of the astronauts managing the mission in space.

The flight simulator and other training was especially critical since Swigert replaced Mattingly on the team only four days before lift off.

The question remained, as it does for any team: How will they perform in a real emergency?

Chapter 35
Outer Space

Lift Off

L ift off on April 11, 1970 from Cape Kennedy (now known as the Kennedy Space Center at Cape Canaveral) went without a hitch. This is no small feat for a twenty-ton spacecraft balanced on the nose of a 363-foot-tall (111 m) Saturn V rocket, filled with over 200,000 gallons (750,000 liters) of kerosene fuel and over 300,000 gallons (1.1 million liters) of liquid oxygen needed for combustion, with approximately twelve million working parts. One tiny malfunction could result in an explosion having the blast force of a small nuclear weapon.

For the first two days, everything went according to plan. After firing and jettisoning two of the Saturn V's three engines, the spacecraft entered a stable "parking orbit" 102 miles (164 km) above the Earth. There, the crew took the time to stow equipment, calibrate instruments, and prepare to leave the proximity of Earth. After one full orbit the third-stage Saturn V rocket ignited, propelling Lovell, Haise and Swigert's spacecraft out of its near-earth orbit and onto a trajectory toward the moon. That third-stage rocket was also successfully jettisoned.

It would take careful control of the ship's momentum to harness the slingshot effect of the moon's gravitational field to bring them home again.

The major components of Apollo 13.
(*The diagram is based on an image provided by NASA.*)

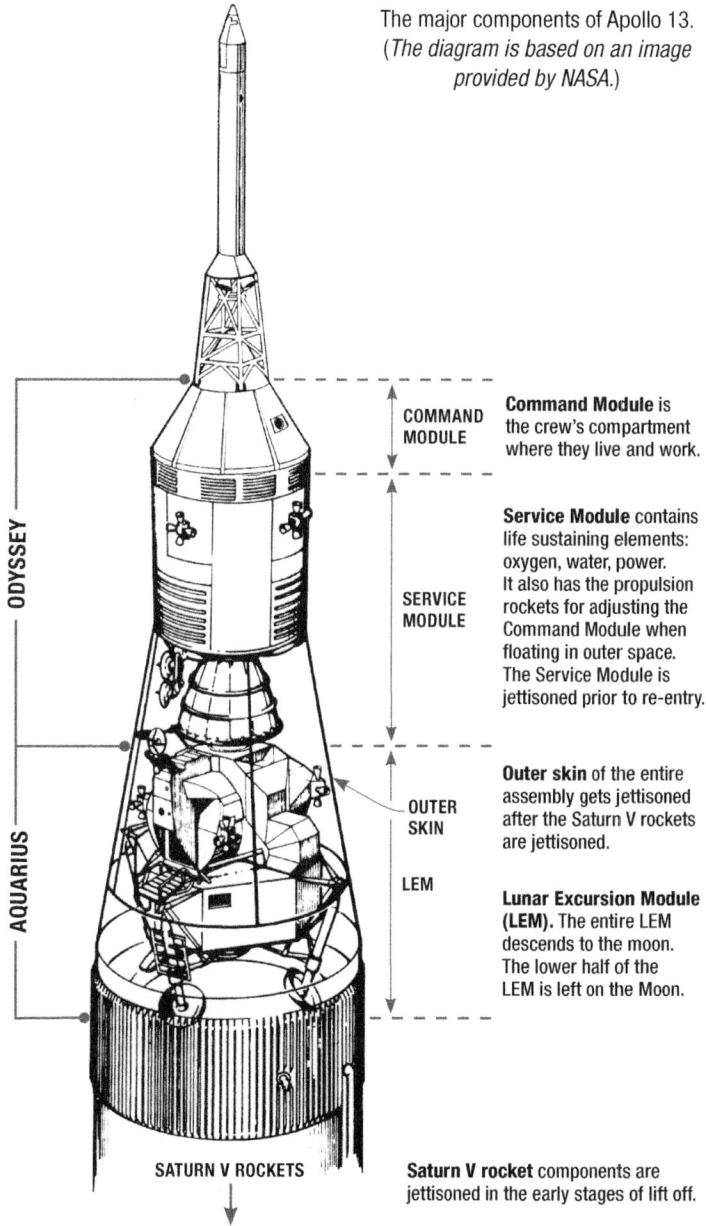

COMMAND MODULE

Command Module is the crew's compartment where they live and work.

SERVICE MODULE

Service Module contains life sustaining elements: oxygen, water, power. It also has the propulsion rockets for adjusting the Command Module when floating in outer space. The Service Module is jettisoned prior to re-entry.

OUTER SKIN

Outer skin of the entire assembly gets jettisoned after the Saturn V rockets are jettisoned.

LEM

Lunar Excursion Module (LEM). The entire LEM descends to the moon. The lower half of the LEM is left on the Moon.

ODYSSEY

AQUARIUS

SATURN V ROCKETS

Saturn V rocket components are jettisoned in the early stages of lift off.

From here on out, there would be no turning back.

The rocket to the moon

The spacecraft was made of three entirely separate components, manufactured by three different aerospace companies, stacked in a compact cylinder atop the Saturn V.

For the aerodynamic efficiency of the launch these were arranged with the conical Command Module – the crew's main compartment – at the top. Next was the cylindrical Service Module containing the rocket engine and its fuel that would propel the assembly while it is in outer space. The Command Module and Service Module combination were given the call sign: Odyssey.

That sat upon the Lunar Excursion Module, the wider, odd-shaped LEM moon lander that would take Lovell and Haise to the surface of the moon once the spacecraft was within the moon's orbit. The LEM's call sign was Aquarius.

For ease of explaining what happened next, we will use the individual component names Command Module, Service Module and LEM, rather than the less descriptive code names Odyssey and Aquarius.

These linked modules, the Command Module, Service Module and LEM, were wrapped in a metal skin to make one tapered and sleek unit for take off, as shown in the diagram.

After lift off and once the larger stages of the Saturn V were cast loose, the spacecraft entered its natural habitat – freely floating in the vacuum of outer space en route to the moon. Aerodynamic efficiency was no longer a concern and now in the vacuum of space the components of the spacecraft had to be rearranged so the Service Module's rocket engine would be at one end, not in the middle of the package. In the new position it could be fired as needed to speed the spacecraft up or slow its momentum down. Once the Command Module and the LEM were properly connected and bolted together into one airtight unit for the rest of the journey, the astronauts could safely pass from one module to the other.

Once arranged in this manner, a burst of power from the Service

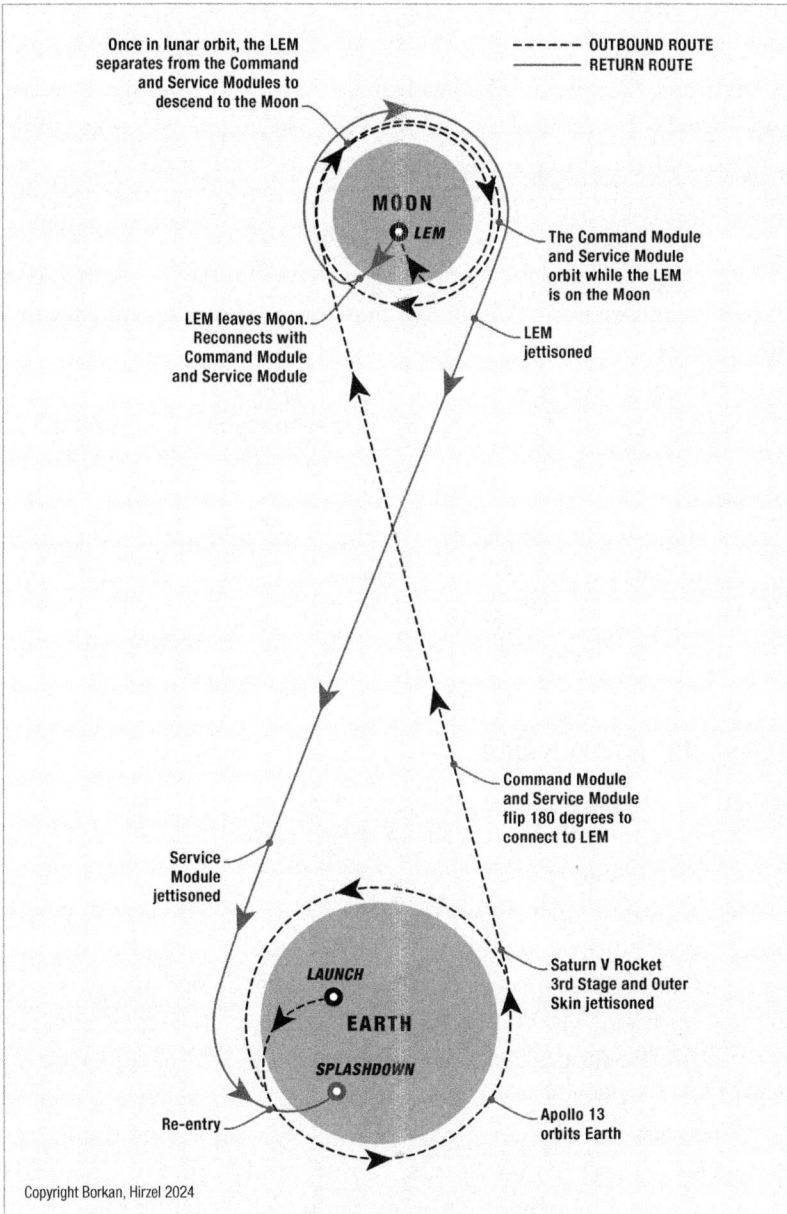

Once in lunar orbit, the LEM separates from the Command and Service Modules to descend to the Moon

OUTBOUND ROUTE
RETURN ROUTE

MOON
LEM

The Command Module and Service Module orbit while the LEM is on the Moon

LEM leaves Moon. Reconnects with Command Module and Service Module

LEM jettisoned

Command Module and Service Module flip 180 degrees to connect to LEM

Service Module jettisoned

Saturn V Rocket 3rd Stage and Outer Skin jettisoned

LAUNCH
EARTH
SPLASHDOWN

Re-entry

Apollo 13 orbits Earth

Copyright Borkan, Hirzel 2024

The intended flight path of Apollo 13.
(*The diagram is based on an image provided by NASA.*)

Module's engine would give Apollo 13 the directional momentum to coast to the moon. Once there, the entire spacecraft would be rotated 180 degrees, engine bell forward, slowing it down enough to enter its planned orbit around the moon. This 180-degree rotation would happen again later in the voyage, to help Apollo 13 speed away from the moon for return to Earth, and again days later, to slow down in preparation for splashdown in the Pacific Ocean.

This complicated dance would have to be flawlessly completed at 20,000 miles per hour (32,100 km per hour). The diagram explains this process.

Complicated, but it was all a part of the plan.

Once completed, the crew would be comfortably ensconced in the Command Module and coasting toward the moon. All they had to do was attend to routine chores, of which there was an innumerable list. One of these was to physically stir the cryogenic tanks containing the liquid oxygen and hydrogen that would fuel the mission for its duration. They were three quarters of the way to the moon when Swigert flipped the switch to begin the stir and, along with the other astronauts, turned his attention to other routine matters.

A devastating explosion

Sixteen seconds later, a *bang-whump-shudder* shook the ship.

An amber light flicked on above their heads.

An alarm began to sound.

Another light began to glow on the electrical instrument panel. The readout showed an abrupt loss of power in "main bus B," one of the two electrical distribution panels that provided all the power to the Command Module. If one bus went completely dead, half the systems in the module in which the astronauts lived and worked would go dead with it.

"Houston, we've had a problem. We've had a main B bus undervolt." Jim Lovell's carefully nurtured ability to keep his emotions out of his voice was being put to the test right now.

Whatever had just happened was now unfolding before them. It was a situation for which there was no precedent. They were hurtling toward the moon with no way to stop, and if the power went down completely, no way to control their spaceship, and no possibility of survival.

Nothing in the team's training had prepared them for this exact technical problem, and yet everything had prepared them for dealing with the unexpected in a cool and controlled manner.

Almost immediately, the indicator for oxygen tank number two dropped to zero. It appeared that one half of the oxygen required for the trip had vanished.

The men and women at Mission Control, led by Chief Flight Director Gene Kranz, read their own instruments from the ground and confirmed the loss. They confirmed that even with one oxygen tank empty, there was enough in the other to provide for the entire mission.

But then, the spacecraft began to wobble uncontrollably. Lovell tried to control the ship manually, but as soon as he had it stabilized, it began to shimmy and wobble again.

Outside the Command Module's diminutive windows, a cloud of gas appeared – thin and shimmering, and already stretching for miles. This was a sure sign of unplanned venting, and just about the last thing any astronaut would want to see. Maybe a meteor had breached the integrity of the ship.

With oxygen tank number two already empty, the gauge for tank number one was showing to be half full and visibly dropping. Once this tank drained dry, air and electricity would have to come from the small reserves intended for the very end of the voyage, when the Command Module would be separated from the Service Module altogether. At that point, the Command Module in which the astronauts would be seated would readied for splashdown if and when they returned to Earth.

But Apollo 13 was still outward bound. There would not be enough to last the four days left until splashdown. And unlike an airplane, an Apollo spacecraft could not just turn around in mid-space and return to Earth.

All this had happened in a matter of minutes, on the third day of a ten-day mission. It was already clear that the moon landing would have to be aborted. If the return to Earth took another five days, the three astronauts would not survive.

Now was the time for all that training to really kick in. There was work to do, and despite having come together only a few days before lift off, they'd have to work expertly as a team to achieve it.

The LEM

The Command Module, the astronauts' main living and working quarters, was for the time being snugly attached to the LEM with its own limited supply of air and electricity. But the LEM was intended only for two men – not three – and only for the few days it would be needed for the actual lunar landing. The idea of using the LEM as a lifeboat had been talked about at NASA, but never trained for, and never tested. By the time Mission Control raised the idea of using the LEM this way, the three astronauts had already come to the same conclusion.

The two-man interior of the LEM was about to become home for three. They would have to "abandon ship" for the lifeboat, and shut down the Command Module completely, to conserve remaining power for the return home.

Haise was the one who knew the LEM best. His dedication to learning its many systems and secrets was about to be put to the test. With only two seats intended for the lunar landing crew, there was no room for three people. An improvised third seat on top of the hatch cover behind the other two would have to do. They would learn fast how to stay out of each other's way.[17]

The remaining battery power of the Command Module, still attached to the LEM, would be enough to control the ship for re-entry to Earth – if no further disasters happened. While Swigert in the Command Module raced through the steps to power it down, Haise and Lovell were busy in the LEM bringing it back to life.

---

[17] In some ways this is reminiscent of situations faced by polar explorers and climbers of Everest. For survival they often had to crowd more team members into a small tent.

Had this disaster happened during the return voyage from the moon, the oxygen tanks in the LEM would have been depleted.

The powered-down Command Module began to rapidly chill in the absolute-zero temperature of outer space, while the LEM was slowly warming up.

If any of the men were fearful they were not about to let their teammates down by showing it. With both tasks complete, they could relax just a little bit and prepare for the next critical maneuver for their crippled spacecraft.

Even though oxygen tanks were depleted and the three men were crammed into the diminutive LEM, they all knew that to get home, they would still have to travel the rest of the journey to the moon, orbit it, and gain enough momentum in the process to propel them Earth-bound.

And that's not all. There was also the exceptionally tricky problem of re-entry into Earth's atmosphere. And if they survived by not bouncing the spaceship off the atmosphere or burning it up upon re-entry, they still had to land it in an ocean close near a waiting ship to pick them up out of the water before the capsule sank.

It would take great leadership, supremely cool nerve, and a team effort to survive this.

Free-return burn

The "free-return burn" from the main rocket engine on the Service Module was standard operating procedure for all the Apollo missions. As the ship drew near to the moon, it must be positioned to enter the moon's gravitational pull in the right place, at the right speed.

Too fast, and it would not be pulled into a lunar orbit. The spacecraft would then fly into the vast reaches of outer space with no possibility of return.

Too slow, and it would be captured into such a low orbit that it would slowly disintegrate. Eventually, the ship would crash to the surface.

The free-return burn was a relatively simple procedure, but it required a star fix for determining the spacecraft's location to fine-tune it. The cloud of gas that had vented from the ship earlier was traveling with them, obscuring all but the brightest stars. With the ship still wobbling, there was simply no way to get that fix. Quick blasts of the small thrusters were normally all that would be required to obtain the necessary fine alignment.

Lovell had done it before, from the Command Module of Apollo 8. Now he would have to do it in half the usual time.

The first step was to rotate the ship 180 degrees, so that the rocket engine bell was pointed toward the moon. Lovell took the controller in hand and gave the thrusters a gentle pulse. The ship lurched into a sudden, violent yaw, almost out of control.

No one had prepared for this moment – the LEM's thrusters were intended to guide only the LEM during a moon landing, not the LEM with the cold, dead carcass of a 63,400-pound (28,750 kg) orbiter stuck to it, giving them an entirely different center of gravity.

The lurch almost threw the guidance gimbals out of alignment, thus denying all capacity for guiding the ship at all.

With no way to get an accurate fix from the ship, the Houston Command Center would have to radio up the coordinates and Lovell would have to punch them into the computer. He'd have to hope the guidance platform was correctly aligned to aim the spacecraft in the right direction to the moon, around the dark side of the moon – where all communication to Houston would be lost – and emerge facing homeward.

If it was successful, the crew would be headed for Earth. If not, they all knew too well what that meant.

Once the ship was properly oriented, there would be three options for power and duration.

The "superfast" would get them back to Earth quickest but force a splashdown in a remote part of the Pacific Ocean where it would take a rescue ship longer to reach them than they had.

The "fast" would get them to a slightly friendlier part of the ocean.

The "slow" burn – least dramatic and with the longest return time – would splash them down in a location where US Navy ships would be waiting, even though it would add almost a full day to their journey.

The wisest choice was the third.

Lovell oriented the spacecraft as best he could under the circumstances, and controlled the burn strength and duration. When it was completed, Houston reported that the ship was aligned for the correct orbit around the moon and the free-range trajectory that would propel it homeward. Once it emerged from the dark side, it would be perfectly aimed for Earth.

This much was a success. With nothing to be done for a few hours, they could relax again and take in the spectacle of the moon from up close that was filling the windows with its mysteriously cratered, grey face.

Everything seemed to be in order.

"O.K. Jim," said Mission Control, "we have a little over two minutes until loss of signal, and everything's looking good here."

"Roger. . . O.K. we'll just sit tight, then. See you on the other side."

For nearly twenty minutes on the dark side, the three men were out of communication with Mission Control, out of touch with Earth, and had nothing but the coldly burning stars for company.

Chapter 36
Outer Space

Critical Team Moments

Apollo 13 emerged from behind the moon on schedule, and sent a jolly, "Good morning, Houston. We're really zooming off now. We're leaving."

Haise and Swigert remained at the windows, eyes fixed on the departing moon, a view seen only by a rare few people. When Lovell reminded them of their duty, they quickly took their seats.

The danger was far from over. It was time to prepare for another critical firing of the rockets: the PC+2 burn.

Positioning the spacecraft would be a little easier now. They had left the gas cloud behind them. It had forged off into space when they'd entered lunar orbit, giving Lovell a better sense of how to handle the thrusters in the LEM. The success of the PC+2 "speed-up" burn would determine whether they would splash down in the South Pacific, Atlantic, or the Indian Ocean.

They would have to use the sun for a navigational fix. It would be, Lovell thought, like aiming for the broad side of a barn door.

Haise repeated the instructions for the 180-degree rotation of the ship to point the rocket engines in the correct position, to make sure he'd heard correctly.

"Kind of an imprecise method, don't you think?" asked Swigert. "Do you have any better ideas?"

Lovell paused. "None at all. Do you?"

"Nope. Let's get started."

To their relief, the improvised maneuver was a resounding success. The PC+2 burn, like the earlier lunar insertion burn, was controlled from the LEM. It worked as before, flawlessly.

The sleep deprived crew then would have to re-establish the "barbecue roll", a maneuver that would prevent overheating on one side of the ship and over-freezing on the other, followed by powering down Command Module to conserve energy.

Once that was completed, Lovell and Swigert retired to the increasingly colder Command Module. The air temperature was 58° F (14.4° C) and falling – for a few hours' long-denied and fitful sleep. Haise stayed in the LEM.

Long before this trip, Haise had earned the reputation of a man not easily rattled. He rather enjoyed the solitude, the elbow room, and the feeling of being in control of his own ship, the LEM part of the spacecraft. When he noticed a loss of helium pressure, Haise made the decision not to wake Lovell. There was nothing to be done about it anyway. They were all in desperate need of sleep because rest period allotments were stingily doled out by Mission Control.

Another problem, another improvised solution.

Houston now turned its attention to the next problem: how to scrub the carbon dioxide from the air in the LEM.

The team had generated five days' worth of the gas by simply breathing, which in turn had reduced the amount of breathable air available. The filters in the LEM were not interchangeable with those in the Command Module. They were smaller, intended to serve only two men for two days, and by now were dangerously overfilled. Failure to resolve this problem would mean asphyxiation for the men long before they could open the hatch doors to fresh air on Earth.

Houston radioed up another improvised solution to another intractable problem. They suggested the team make the filters from the Command Module work for the LEM by cobbling together a case for them out of bits of cardboard, and flex hose held together with tape.

This, like all the improvised solutions thus far, worked. The CO_2 levels began to drop immediately.

A little more rest came, but it was not pleasant in the ever-dropping interior temperature. But to save the minimal remaining battery power for the last necessary maneuvers before re-entry, the heater could not be used. Temperatures in the space craft had fallen to 38° F (3° C).

They then needed to execute a series of procedures, the first of which was to rotate the ship one more time, to get the rocket engine bell pointed toward the earth, to slow the spacecraft down for re-entry. This could only be achieved by partially powering up the cold, dead Command Module again with the minimal electricity remaining in its batteries. This had never been attempted before. Mission Control came up with a three-page sequence of steps that must be taken, in turn, and to the exact degree. Unlike every activation procedure Swigert had ever followed in training, this one was *wholly* improvised.

Several of the many people at Mission Control tasked with bringing Apollo 13 home. Standing closest to the left is Ken Mattingly who was scrubbed from Apollo 13 due to risk of illness. *(Photograph: NASA)*

Houston radioed up the individual steps one by one. Swigert dutifully, and accurately, wrote them down. Failure at this juncture so close to Earth when survival seemed almost possible, would doom the men and the mission all over again.

The partial power-up and the ship rotation worked.

With the rocket engines now pointed toward Earth, it was time to ignite the final burn to fire the engines enabling them to leave the free-return trajectory.

"All right," Lovell said, "Jack [Swigert], since we don't have any countdown clock, you time the burn with your watch. We're firing for 14 seconds at 10 percent. Freddo [Haise], since we don't have an autopilot, you grab your attitude controller and keep us from yawing too much. I'll handle pitch and roll with my controller and also take care of ignition and shutdown. Got it?"

Haise and Swigert nodded. This critical procedure was a team effort.

The few seconds passed.

"Houston, burn complete," Lovell said.

The Command Module was powered down again until it would be fully awakened just before re-entry.

Goodbye, LEM

Safely on the correct return trajectory, the next task was to jettison the Service Module. This was completed. As it drifted away, slowly rotating, the men were able for the first time to see the cause of that original outward-bound *bang-whump-shudder*. The panel covering oxygen tank number 2 was entirely gone; there was only a charred, empty cavity in its place. The tank had blown up and doomed the mission. It was still not clear what was left of Command Module would survive re-entry. Had the explosion damaged the Command Module's heat shield?

Even a hairline crack in the epoxy would mean certain failure on re-entry into Earth's atmosphere. Without a fully functioning, stable heat shield, they would be cooked alive in the 2,700° F (1,480° C) oven that the capsule would become.

Haise's precious LEM, that had become their indispensable lifeboat, was jettisoned next. Had the mission been successful, part of it would now be an artifact on the moon's surface, an eternal testament to their presence. Now only the Command Module with three men in it remained.

Their final approach for re-entry would require a full power-up of the Command Module. Would all the conservation of battery power that had caused so much cold and sleepless misery for the team provide enough amps for this last critical maneuver?

Ken Mattingly, the astronaut who should have been on Apollo 13 with Lovell and Haise but was scrubbed from the mission due to having been exposed to German measles a few days before lift off, played an instrumental role in working with Mission Control. Mattingly radioed up the new plan: a full power-up of the completely cold Command Module. It had never been attempted before. The complex instructions from Mattingly to the astronauts took two hours to convey.

The astronauts successfully executed the plan. The Command Module powered-up.

All was set for re-entry now.

All that was left of the Apollo 13 mission was the Command Module, with its heat shield and drogue parachutes, and the very small team within it awaiting their fate.

There was nothing more to be done. Even if there had been, there was no power left to do it with.

Home again

In a spirit of international cooperation, given the uncertainty of the location of Apollo 13's ocean-based landing, nations from all over the globe offered their ships to pick up the astronauts.

Apollo 13's Command Module splashed down in the Pacific Ocean near Samoa at 12:07 p.m. on April 17, nearly six full days after its launch.

The three astronauts were picked up by the USS *Iwo Jima*.

They had made it home.

Once they'd become accustomed to the fact that their mission had rapidly evolved from a moon landing that they had worked so hard to earn, to one of pure survival against insuperable odds, they simply got to work on their new calling. Their training, resourcefulness, teamwork and their steely, unflappable nerve had saved them.

This very small team accomplished the most remarkable recovery from overwhelming adversity, but they could not have done it without a large team supporting them – Gene Kranz, Ken Mattingly and the many men and women at Mission Control in Houston who exhibited the same dedication and resourcefulness as the astronauts did.

Aftermath

None of the three men ever went into space again. The twenty planned Apollo missions were cut to eighteen by a newly elected, parsimonious presidential administration. Lovell, Haise and Swigert all remained with NASA in advisory positions for a time. After all, who knew better

The crewmembers of the Apollo 13 mission step aboard the USS *Iwo Jima*, following splashdown and recovery operations in the South Pacific Ocean. Exiting the helicopter which made the pick-up are *(from left)* Haise, Lovell and Swigert. *(Photograph: NASA)*

how to deal with the unforeseen in space than these three?

Haise stayed on with NASA to test-pilot landing procedures for the new Space Shuttle, the next-generation workhorse under development. Lovell and Swigert left NASA for other pursuits in private industry, including for each a foray into politics. The three men met from time to time for special NASA commemorative events, flying in from their respective homes and new careers, but their once-bright dreams of landing on the moon would never be realized.

As a small team they proved something valuable. Dedication to a new goal, cohesiveness of spirit, remaining unflappable in the face of real and imminent danger, and high levels of teamwork could achieve the near impossible.

Had their mission gone without a flaw, Lovell, Haise and Swigert would be known only by those with an interest in the Apollo missions which landed men on the moon. Thanks to their brave actions, commemorated in the Academy Award winning Hollywood movie *Apollo 13,* their legendary teamwork will be known by generations of people to come.

Also, the calmly stated phrase, "Houston, we've had a problem," has become firmly embedded into our popular culture.

★

As we were writing this chapter, NASA Artemis missions to the moon are being readied for launch. Artemis I was an unmanned flight that flew to the moon but did not land. It was a test in preparation for manned Artemis missions. While it successfully completed its journey, some of the heat shields had broken off.

Artemis II will go around the moon but will not land – similar to Jim Lovell's first Apollo 8 mission. The four person crew led by Reid Wiseman includes a woman Christina Koch, a Black man Victor Glover, and the Canadian astronaut Jeremy Hansen. The following year, Artemis III will land on the moon enabling two of the four person crew to walk on the moon's surface. One of the lunar landing party will be a woman.

However, during the final edits of this chapter, NASA just announced both launches will be delayed, with Artemis II pushed back to September 2025 and Artemis III to September 2026. Sending crews to the moon is as challenging in the hyper-computerized age of the 21st century as it was in the under-resourced age of the 20th century.

★

The Apollo 13 team survived disaster in the cold expanse of outer space. The final team we'll look at is one that faced disaster in another cold expanse: Antarctica. In their case, they had no "Houston" to call.

Danger surrounded them. Would they survive?

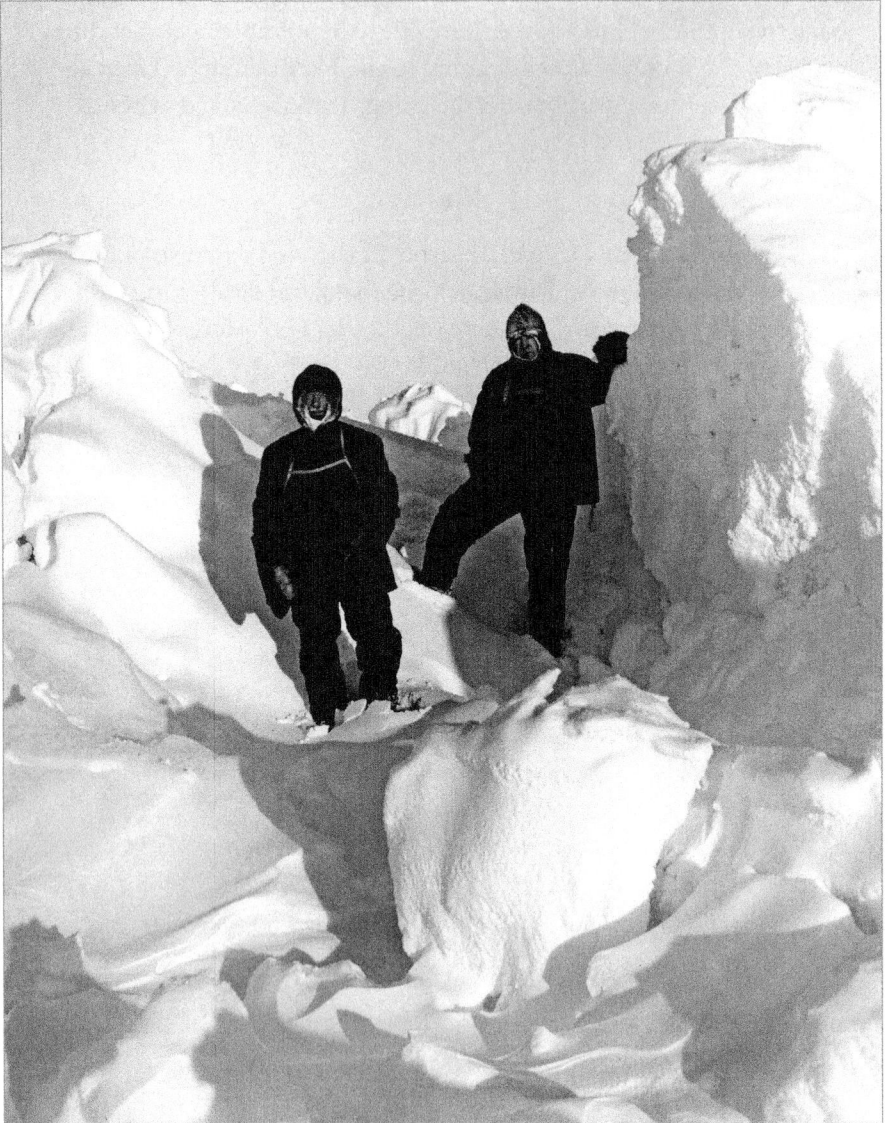

Frank Wild *(on the left)* and Ernest Shackleton standing near ice hummocks during the *Endurance* Expedition 1914-1916. *(Photographer: Frank Hurley)* *(© Royal Geographical Society [with IBG])*

Chapter 37
Survival

Shackleton and Wild

"He is my second self. I love him, as does every decent man on the expedition. He has been a tower of strength to me..."

—*Excerpt from a letter that Ernest Shackleton sent to a friend.*

Ernest Shackleton and Frank Wild spent more time together in the Antarctic cold – struggling up glaciers and down again, shipwrecked on the sea ice, and in the end up together on board the *Quest* in search of new areas to conquer. They shared these experiences with other men on the expeditions, but a deeper bond developed between them – one that exceeded the ordinary realm of a friendship that extended over two decades.

Different ranks, same expedition

Shackleton and Wild first met the moment they both joined Captain Robert Scott's 1901-1904 British expedition to Antarctica on the ship *Discovery*.

Unlike most of the other Royal Navy men chosen for that first major exploration and scientific expedition into the Antarctic, Shackleton and Wild had each come up through the ranks. Both earned their skills the hard way – in the British Merchant Navy on cargo and passenger ships. Shackleton joined the *Discovery* Expedition as third officer (the only non-Royal Navy officer), and Wild joined as an AB (able-bodied seaman).

Neither Shackleton nor Wild were a perfect fit for Scott's *Discovery* Expedition, a civilian enterprise that would be run under naval principles and discipline. Shackleton, lower in rank than any of the Royal Navy officers on the expedition, could only gain admittance to the "club" by enrolling as Royal Navy Reserve, then find a place in the expedition where he could start to make a name for himself.

Frank Wild, with all his seafaring experience, had only recently enlisted in the Royal Navy. Shackleton and Wild's bond, formed first out of mutual respect during the thirteen-week voyage from England to Antarctica. Their friendship took root and grew at a steady pace during the voyage even though they were separated by rank in the traditional Royal Navy way of "officers" and "men." This distinction was a class distinction and not a gender-based one, since all the early Antarctic explorers, scientists, and crews were men. The first women to over-winter in Antarctica did so in 1947.

Shackleton was in the officer corps; Wild was with the "men."

First, Discovery

Scott's *Discovery* Expedition was intended to place Britain in the vanguard of Antarctic exploring nations. Since other nations – Sweden and Germany in particular – were also vying for such visibility in 1901, Britain went all-out to make the most significant marks on the map of the as-yet-unexplored continent.

Scott's plan was to land a large party as far south as the Discovery could get, given the sea ice conditions, and build a hut on shore for use by the scientific staff. The entire team would use the ship as their sleeping quarters over the winter. In the Antarctic spring, field parties would explore the Western Mountains and seek out the South Magnetic Pole[18], while Scott would lead a three-man party on a trek

---

[18] The South Magnetic Pole is different from the South Pole. The South Magnetic Pole is the one

far into the interior, at least halfway to the South Pole.

No one had ever attempted a Polar expedition so ambitious. Captain Scott, along with practically everyone else involved, had no polar experience. The overall parameters of what would be required in this extreme environment—supplies, how many men would travel, and who should lead them – were derived from outdated standards set by the Royal Navy decades earlier, before Antarctica was even discovered. They were based on Sir John Franklin's 1845 naval expedition to find the Northwest Passage (an elusive sea route across Northern Canada connecting Europe to Asia). That expedition was a dismal failure, with two ships and their 129 men permanently lost to the cold and ice.

Scott's *Discovery* Expedition intended to set a new example with better rations and equipment, and more inspired leadership. The best men for the *Discovery* Expedition would be the most adaptable, those who could enter an unknown situation and make the best use of their innate strengths.

By this definition, two of Scott's best men were his third officer, Ernest Shackleton, and one of the able-bodied seamen, Frank Wild.

Shackleton and Wild

Once in Antarctica, Wild distinguished himself almost immediately.

An overland party went out on a multi-day mission to place a message post at Cape Crozier[19], 35 miles (56 km) from the newly established *Discovery* camp at Hut Point, on Ross Island. In the days before radio contact, an expedition ship would have to post messages in designated locations for scheduled relief ships to find.

On this trek, no one was prepared for the weather.

When a blizzard swept down on the ten-man party, the Royal Navy officers in charge sent seven men back to the ship at Hut

---

place in the entire Southern Hemisphere where a magnetic dip needle would point down in a straight vertical line.

[19] Cape Crozier was the destination of Wilson, Bowers and Cherry-Garrard in their quest for emperor penguin eggs, as described in chapters 11-13.

Point. In the blinding white-out of the whirling snow, the returning seven-man party got hopelessly lost and had no idea how to save themselves. One man fell off a cliff and died. Aimless panic ensued, but the level-headed Frank Wild took control of the situation, calmed everyone down, and led the men first to a safer location then back to the ship.

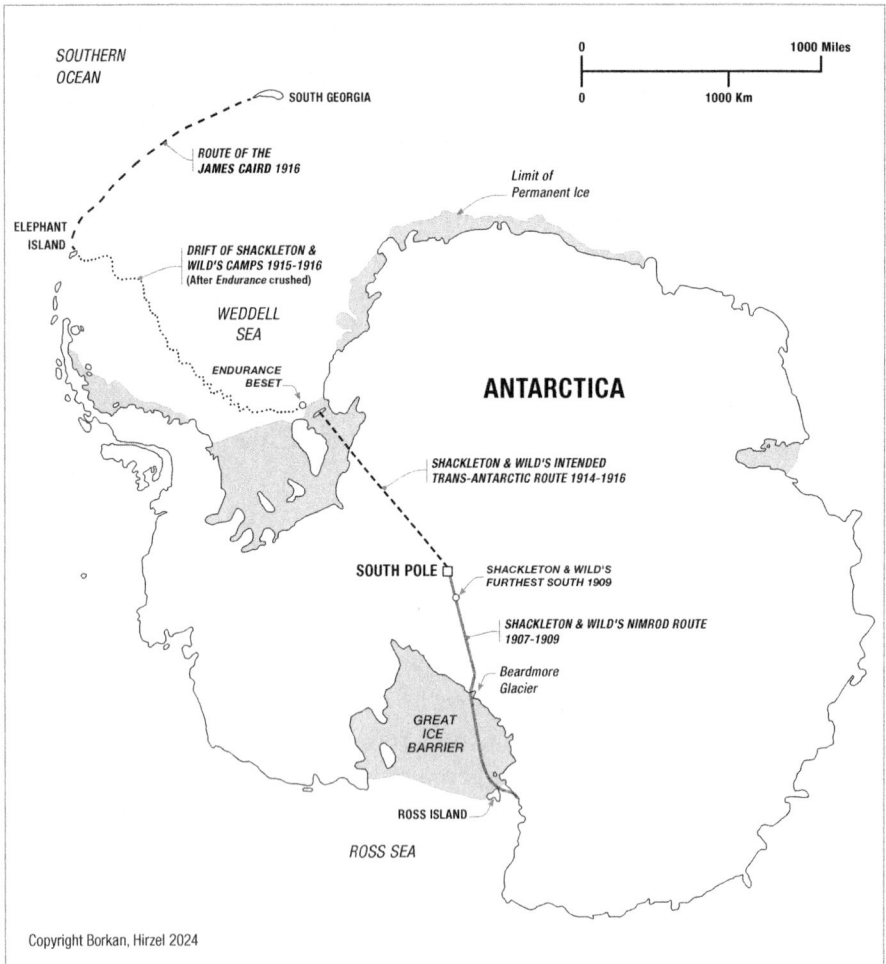

Routes of Shackleton and Wild's *Nimrod* and *Endurance* Expeditions.
(Locations and distances are approximate.)

These traits stood him in good stead throughout his long Antarctic career – first as an AB on the *Discovery* Expedition (1901-1904), then as an officer in Shackleton's *Nimrod* Expedition (1907-1908), a field leader in Mawson's Australian Antarctic Expedition (1912-1914), and ultimately as Shackleton's second-in-command on the *Endurance* (1914-1916) and *Quest* (1921-1922) Expeditions.

The *Discovery* was intentionally frozen in near Hut Point for two years before the sea ice broke up and released the ship for her voyage home in 1904. Shackleton was the only officer sent home on the relief ship after the first year in Antarctica, along with almost all the merchant sailors who did not quite fit in with Scott's naval discipline. The proximate cause of Shackleton's dismissal was the breakdown of his health on the long journey south with Scott and Dr. Edward Wilson. The three of them were the first to trek hundreds of miles into the interior of Antarctica in the direction of the South Pole. Even though Shackleton had contracted scurvy and very nearly died on the return leg of that trek, he took it as an affront to be invalided home.

Shackleton resolved to return to Antarctica in command of his own expedition, to make good on his shortcomings on the *Discovery* Expedition.

Always a good self-promoter, Shackleton found backers for another expedition upon his return to Britain. By 1907 he had just enough funds in hand to buy a ship called the *Nimrod*, fill her with stores and a prefabricated hut, and head back to Antarctica. His plan was to be the first to reach the geographic South Pole – that one place where all lines of longitude converge and is the point of the axis around which the earth revolves. The man who reached that almost mystical place would bring glory to the King, the British nation (though Shackleton was Irish by birth), and himself.

Chapter 38
Survival

Shackleton Picks His Man

S hackleton had taken note of Wild's skill and endurance during their year together on the *Discovery* Expedition. Given the customary division of rank on that voyage, the two could not form a friendship between equals, but Frank Wild was one of the very first men Shackleton enlisted for his *Nimrod* Expedition. Since Shackleton was now in complete charge, he would manage the personnel in his own casual, inspired manner.

It was on this expedition that Shackleton and Wild bond cemented.

Shackleton ran the *Nimrod* Expedition in a more egalitarian manner than the standard Royal Navy way. The shore party consisted of fourteen men living in a 19 foot by 33 foot (5.7 m by 10 m) wooden hut they built at Cape Royds on Ross Island, 20 miles (32 km) north of Scott's *Discovery* Expedition's scientific hut. The men who stayed on shore paired up and built cubicles, each 6 feet by 7 feet (1.8 by 2.1 m) within the hut for themselves. These would house two men in each, their beds and all their gear.

Frank Wild paired up with Ernest Joyce, another *Discovery* Expedition pal. Their cubicle also housed a printing press. They named their cluttered lair *The Rogues' Retreat* and made themselves at home until winter's end marked the beginning of the *Nimrod* Expedition's great southern push, with the goal of reaching the South Pole.

Shackleton, as leader, was the only one with his own private cubicle. No great distinction was made between the officers, the scientists, and the men when it came to the work of the expedition, and there was a lot of work to be done.

Although the discovery of the South Pole was the primary goal of the expedition, pinpointing the exact location of the South Magnetic Pole was equally ambitious. This, like the geographical South Pole, was over 800 miles (1,287 km) from the winter quarters at Cape Royds, but in a different direction. For each, the round trip out and back meant that the explorers would trek over 1,600 miles (2,575 km), dragging all their food and gear every step of the way. These were the journeys that would bring worldwide acclaim to the expedition, and especially to Shackleton.

These were extremely dangerous ventures.

The parties were small. Three men (Douglas Mawson, Edgeworth David, and Alistair Mackay) for the South Magnetic Pole Party – and four (Shackleton, Frank Wild, James Adams, and Eric Marshall) for the South Pole. The distances to be covered would take months.

The routes to each led over vast fields of crevasses that could open underfoot without warning, dropping the unlucky man to the length of his sledging harness. The parties were not able to carry enough food for the length of time it would take to walk that great distance out and back. Depots of food and supplies would have been placed partway along the route before the start, with the intention that they would be dug out of the drifts of snow on the return leg of the journey. In addition, as the ponies that were used to help pull the sledges died, their meat would also have been depoted along the way.

Through endurance we conquer

With only four men chosen for the trek to the South Pole, everyone put out his best for the expedition and, often unasked, did more than their very best. Frank Wild was one of those men. They left the base camp at Cape Royds on Ross Island on October 29, 1908. There would be no supporting parties to help them with the sledges heavily laden with tents, sleeping bags, food and cookers – the bare necessities to live in the barren icefields of the Antarctic. When four of their eight ponies intended to move supplies along what they dubbed the South Polar Trail died before the start, the men would have to take on the task of beasts of burden in addition to their intended roles.

On the trail, Shackleton's egalitarian ideas included the switching around of partners for the two-man tents. This ensured that each man had another new fellow to share not only the sleeping quarters but all the sundry household chores that came with it.

At the end of the daily 15 mile (24 km) march, each two-man team would unpack and erect their tent in the unending wind, hand in the sleeping bags, fire up the cooker for a quick hot meal, then clean up and sidle into their nearly frozen fur sleeping bags. Then in the morning, they would do it all over again in reverse order. Each man had his talents and would, by order of Shackleton's egalitarian ideas, have to learn to get along with a new man's shortcomings every week.

Had the teams not been mixed up on a regular basis, petty annoyances could have assumed gargantuan dimensions, or the whole group could have split into two-man alliances, to the detriment of the expedition's ultimate goal. Equally important in these rather intimate circumstances was that each man came to know the others better. Friendships bloomed but could not to grow into cliques. And Shackleton got to know the strengths and weaknesses of his men.

It was during their shared time as tent mates that Shackleton and Wild's friendship flourished. Both men kept diaries. Shackleton's was later edited into his book *The Heart of the Antarctic*. Wild's remained private, but his daily entries revealed his growing admiration for the boss.

The shared work became ever more strenuous for the four men as the expedition extended itself farther to the south. The first 375 miles (600 km) over the relatively level surface of the Great Ice Barrier (today known as the Ross Ice Shelf), before the last of the ponies died were relatively easy. But the next 125 miles (200 km) were not.

The snowy and icy route ahead would take them up the breathtaking Beardmore Glacier, the longest and highest known in the world. The uphill surface was steep, treacherous, and everywhere riven by enormous crevasses. They would have to pull the sledges to the top of the Beardmore Glacier, ascending 10,000 feet (3,050 m) in elevation.

One of the crevasses near the beginning of this long uphill trek swallowed the last surviving pony, Socks. This was a double tragedy. They not only lost the animal's help hauling the remaining two sledges but the meat on its bones. Socks was always intended to become part of their overall rations depoted for the return journey; now that part of their food lay at the bottom of a crevasse, never to be recovered.

As the days wore on for the South Polar team of Shackleton, Wild, Adams and Marshall, the elusive goal seemed to slip away. The men ate less each day, to make their rations last long enough to complete the journey. By choice, they were slowly starving themselves, draining away the energy they would need to make it home. Their bodies never properly healed from the damage wrought by the trail caused by snow blindness, the falls, the accidental wounds, and the cold and near frostbite on their faces and limbs. Shackleton's family motto, *Fortitudine Vincimus* – through endurance we conquer – was a motto for all of them.

From there on the struggles became intense, and the tensions between the men began to mount. Their daily mileage decreased from the absolute minimum of 17 miles (27 km) per day needed to make it to the South Pole and back to Cape Royds in time to meet the relief ship *Morning* before her scheduled departure for New Zealand in March. The ship's departure could not be delayed, as the risk of being frozen in increased with each passing Autumn day. Timetables were strictly kept, even if it meant stranding explorers for another year in Antarctica.

Forward or back

Well out on the glacier, each of the four men, now bound together in a single harness and dragging their heavy-laden sledge up the slippery slope, became acutely aware of who was pulling their share and who was not. The men's diaries include not only details of the new vistas constantly opening before them, but also bitter diatribes against their fellows who failed to pull their hardest.

They continued to swap tent mates every week. A look into Wild's and Shackleton's diaries revealed their mutual admiration. Each had a deepening respect for the other, and a growing disdain for the other two men.

By the time the four men reached the top of the Beardmore Glacier, after three weeks of constant uphill climbing on reduced rations, they were definitely weakening. Although painfully aware that they had fallen irrevocably behind in the schedule, none of them would admit to it. If only the headwind would stop, and they could further reduce their daily ration, they might – barely – make their goal.

Their Christmas dinner was the only reward, a lone opportunity to have a full belly. But what they ate was not enough to recover their strength. They were on the polar plateau, and the way to the South Pole was almost level and free of crevasses, but the 10,000-feet (3,050 m) altitude had sapped all their strengths. It was becoming painfully clear that lightening their load to the bare minimum would not help. They would have to turn back well before reaching the goal they had suffered through so much to gain.

Wild wrote in his diary on January 4, 1909, "By noon we had only done six miles, and we were all done up, we could not keep warm, and found we could do no better with our seventy lbs [loaded sledge weight measured in pounds, per man] than we did a fortnight ago with two hundred, which shows that we have all weakened on shortened food and low temperatures. Last night Doc [Eric Marshall] took, or rather tried to take our temperatures, but his clinical thermometer was not marked low enough to take any except mine. The other three were therefore below 94.2° [F.] which spells death at home."

They erected their last outward camp on January 8. They were 103 miles (165 km) from their goal. Based on the calculations Shackleton, Wild, Adams and Marshall did, it was clear this was to be their turning back point. To continue forward to reach the South Pole would mean death by starvation on the return journey.

They faced a binary decision: go forward or go back. It was here that Shackleton made a remarkable decision. He saw a third alternative.

On January 9, they would leave the tents, sleeping bags and all equipment behind and march south as far as they could for one day. They would plant the British flag, and then turn around and go back to their encampment. On January 10, they would start the journey home.

Leaving behind everything but the bare minimum they planted the British flag a mere 97 statute miles (156 km) from the Pole. Wind and weather, starvation and fatigue had done them in, but on that date in 1909, they had come closer than any humans had toward either Pole at the very ends of the Earth. (Peary and Henson's 1905-1906 expedition turned back about 150 miles (241 km) from the North Pole.)

Shackleton wrote in his diary after the day's work on January 9, "Our last day outwards. We have shot our bolt, and the tale is latitude 89° 23' south, 162° east." This would be their record, but their attempt "within 100 miles" became a talking point later when Shackleton was out raising funds for his next Antarctic expedition.

Shackleton wrote to his wife, Emily, "I thought you would rather have a live donkey than a dead lion."

The return journey

Now began the desperate struggle to make it back to Cape Royds on the starvation rations they had allowed for the return trip. It would be a struggle that the four men would barely survive.

The first few days, with the wind now behind them, they made mileages better than expected, but these early successes did not last long. Along the way out they had left caches of food and cooking oil to pick up on their return, but the spacing of those depots had been set assuming a rate of travel that could be kept by healthy, fit, well-fed men. This they were not.

Spinning out their rations over a period of months, taking only two-thirds of what Shackleton had originally planned for, had taken a deadly toll. Twenty pounds (9 kg) of biscuit for four men over a distance of 140 miles (225 km) was not enough. In reality, for the energy expended, they should each have been eating the equivalent of three Christmas dinners every day.

Once they had descended the Beardmore Glacier and were at the sea level Barrier, they were able to locate depots they left on the outward journey which included the butchered remains of the ponies that had died, one by one, on the way out. Rather than providing additional sustenance, the barely cooked meat gave them all dysentery.

The frequent stops that were now necessary severely cut down on the daily mileage, which in turn forced them to cut down even more on their daily biscuit ration, which weakened them even further. There was no opportunity to rest and recover. A rest and a proper feed would have done them all a world of good, but they had to forge onward as best they could, only stopping to camp when they had reached a state of utter collapse. They were falling very far behind.

Frank Wild suffered worse than any of them, but with every man on starvation rations, there was nothing further to spare for him. His situation was desperate. While breaking camp on January 31, 1909, Shackleton surreptitiously slipped his own last biscuit into Wild's pocket.

"Shackleton forced upon me his one breakfast biscuit and would have given me another tonight had I allowed him. I do not suppose that anyone else in the world can thoroughly realize how much generosity and sympathy was shown by this. I do and by God I shall never forget it. Thousands of pounds would not have bought that one biscuit."

After a very hard struggle, they eventually reached the first large depot set out on the Barrier at the beginning of the journey, on February 20. "Arrived here at 4 pm, pitched tents, dug up the stores, and started in for a meal. We had a full cup of thick and fat pemmican, the best we have had since Xmas, then a black currant pudding . . . finishing up with a cup of cocoa, nearly full strength, and a biscuit." A fine meal and a welcome respite, but not enough to recover from the damage their long starvation diet had inflicted upon them.

The final blow came when Marshall succumbed entirely.

Only 30 miles (48 km) remained to reach the safety of the hut left by Scott's *Discovery* Expedition years before, where a relief team was expected to be. Leaving Marshall and Adams behind in their tent on the Barrier, Wild and Shackleton set out with the remaining sledge, one tent, their sleeping bags, and a minimum of food. If their desperate gamble paid off, they would come back with a relief party from the base camp to pick them up. If they did not make it, then their lives and those of the men left behind would be doomed.

They did make it. In the end all four men reached the hut, and came aboard the *Nimrod*, which had only just arrived at Hut Point. They had achieved one of the greatest polar journeys ever made, at a cost few of their contemporaries would ever fully understand.

And they proved a valuable point: It is far better to "plant your flag" and turnaround from your goal and live to see another day so you can try again, than to "succeed" only to die on the return journey.

A team forged on the ice

The shared hardships and victories between Shackleton and Wild cemented the friendship that would endure the rest of their lives. On the ship back to New Zealand, Shackleton asked Wild if he would join him for another go at the Pole. Even though Wild had written in his diary, "This trip has completely cured me of any desire for more Polar exploration," their epic journey together would only be the first of three.

The two men had become a powerful team with boundless polar ambitions.

Chapter 39
Survival

"Proceed"

F ar from being "cured," Wild went south two years later on Douglas
    Mawson's 1911-1914 Australian Antarctic Expedition. By that time
    Frank Wild was one of the most experienced veterans of the
"white warfare" and had full charge of that expedition's Western Party.

Wild and the seven men under his command were dropped off on
the never-explored coast of Queen Mary Land, 1,500 miles (2,400 km)
west of the expedition's main base. For six weeks, Wild led exploration
teams into the interior of that frozen land, until the ship *Aurora* came
back to pick them up and return the expedition to Australia.

It was in Australia that they learned that the Norwegian explorer,
Roald Amundsen, had reached the South Pole on December 14,
1911 and that Captain Scott, who had been leading another British
expedition to reach the same Pole, had not made it back. News of
Scott's final days and death on the Barrier had yet to reach the outside
world. Amundsen's triumph was undisputed.

The South Pole was no longer a mysterious, undiscovered place and
because of it, had lost much of its nationalistic allure.

But there remained another potential goal.

Except for the route to and from the South Pole from the Ross Sea,
the vast majority of the continent remained unexplored. It had yet to
be crossed from one side to the other, and already there were feelers
out in Germany and Scotland to mount new expeditions with that
goal. Shackleton seized on the idea and began looking for seed money

to finance it. He knew who he wanted on his team – Frank Wild, a proven leader in Antarctic exploration and one of the most experienced of all the veterans in the field.

Soon after Wild's return to England in the summer of 1913, Shackleton enlisted him, opened an office in London, and went public with his plan for the Imperial Trans-Antarctic Expedition six months later.

The plan was to sail a ship through the Weddell Sea, to the recently discovered Vahsel Bay, near the side of Antarctica closest to South America. There Shackleton, Wild and the rest of the six-man team would disembark. The ship would sail back to Britain, while the men on shore would establish a camp in which to overwinter.

In the spring, the six men with six sledges (some of which were motorized; at least one was powered by an engine and propeller from an airplane acquired from Mawson's earlier expedition) and six teams of dogs would head into this part of the unexplored interior, bound for the South Pole. Once they had reached the Pole they would continue onwards, following the route Shackleton's *Nimrod* Expedition had pioneered, toward the known geography of the Beardmore Glacier, down to the Barrier and then to Ross Island.

A separate party would have landed there already. It was called The Ross Sea party. Those men would make use of one of the expedition huts from the *Discovery* and *Nimrod* Expeditions, as well as Scott's ill-fated, final *Terra Nova* Expedition. After overwintering, they would lay out a line of supply depots for Shackleton's trans-Antarctic party to pick up along the Beardmore Glacier and then on the Barrier, on their way south from the pole.

The Imperial Trans-Antarctic Expedition takes shape

While Wild stayed in the office on New Burlington Street, London, sorting out the endless lists of supplies and enthusiastic, frequently inappropriate, applications from aspiring expedition members, Shackleton hit the road trying to attract backers to his new expedition.

The anticipated funding proved elusive. The expedition would end up being run on a shoestring budget, but by spring 1914, there was enough money to buy a 350-ton ship for the Weddell Sea side and rename her *Endurance*. Mawson's *Aurora*, still in Australia, would become the ship to transport the Ross Sea shore party to Antarctica and bring them and the successful trans-Antarctic party safely back to world acclaim.

That was the plan.

The clouds of war were gathering as the *Endurance* was preparing to depart in early August 1914.

When England issued a general mobilization on August 1, Shackleton offered the ship and its crew to the Admiralty for the war effort, but Churchill cabled back a one-word answer: "Proceed."

So they did. The ship would go first to Buenos Aires, and then to the Antarctic island of South Georgia, and on into the ice-choked waters of the Weddell Sea.

Wild was not on the *Endurance* for that first leg. He had taken on the role of dog-management, bringing the sledge dogs as far as Buenos Aires on a separate ship. His experience on Scott's *Discovery* Expedition had taught him how *not* to handle sledge dogs, and he knew from Amundsen's success that properly cared for and driven dogs would be a great improvement over manhauling.

Shackleton came down on a separate ship, and there was much to do in a hurry when they boarded the *Endurance*. While Shackleton pursued last minute purchases and negotiations, Wild oversaw the work on board: building kennels for the sixty-nine dogs, stowing tons of supplies and equipment, and ensuring that the *Endurance* was as ready as she could be for the work ahead.

The *Endurance* landed at the whaling station at Grytviken in South Georgia. The local whalers who went out daily in steam catchers to make their catch all came back with the same report: the Weddell Sea would have more ice than previous years, making travel there extremely difficult.

South on the *Endurance*

All the same, the calendar demanded a start. There was a long distance to go to the desired landing place at Vahsel Bay, a shore party to be landed, a hut to be built, and some advance depots to be laid toward the Pole – all before the Antarctic autumn closed in. Shackleton, a man to trust in luck when the odds were against his plan, decided to proceed.

As predicted, they encountered the ice soon after departure from South Georgia. It was indeed dense and hindered their way, but it did not stop southward progress entirely. They saw a stretch of previously unknown land and named it the Caird Coast, after the expedition's major benefactor. There was a possible landing place here, 200 miles (321 km) north of Vahsel Bay. Such a distance would add at least ten days of overland travel to the polar party's journey.

Shackleton's gamble on getting all the way through to Vahsel Bay did not pay off.

The ice closed in on the ship when she was still 60 miles (96 km) away. When it became clear that the *Endurance* was trapped in the ice and there would be no further progress to the south, Shackleton deemed the *Endurance* a "shore station". He housed the dogs in makeshift igloo kennels on the ice floe, and had the main hold emptied out and converted into "the Ritz," a cozy winter quarters for the officers. The sailors remained in their fo'c's'le, Shackleton in his own cabin. Wild joined three others – second officer Tom Crean and George Marston (both experienced Antarctic men). Captain Frank Worsley was in the wardroom cabin on deck.

The ship, immobilized by the ice, slowly drifted counterclockwise with the large ice floe, through the Weddell Sea. It was an imperceptible motion that carried the ship ever farther away from land. There would be no landing and no crossing of the continent this year. But the motion indicated that eventually the ship would be carried northward to warmer open water where the ice floe would break up and the *Endurance* could return to South Georgia, to try for the Pole again the following year.

Abandon ship

That was not to be.

After five months of drifting, the ice, in its slowly revolving turmoil, closed in ever more tightly. On October 27, 1915, it crushed the ship. The surrounding ice was solid. The nearest known land was at Snow Hill Island, 346 miles (557 km) away. They realized that even exerting their best efforts, it was too far to reach.

The men moved from the stricken ship to live in tents on the broad ice floe. They were able to offload everything they would need to live on the ice for months. Shackleton named this Ocean Camp.

The men kept three of the *Endurance* lifeboats, knowing the boats would be needed when their ice floe eventually drifted to the warmer water to the north and broke up under them. After an abortive attempt to drag the lifeboats over the floe towards its edge, which almost resulted in a near mutiny by two men that was swiftly shut down by Shackleton and Wild, the camp relocated to another part of the large floe.

Shackleton named it Patience Camp. All they could do was have patience and wait.

During all this time, the steady voice of Frank Wild as Shackleton's right-hand man helped to calm and bring order among the men. Shackleton, known as "The Boss," made daily visits to each tent to enquire after the men's health. This helped to maintain a sense of order if not tranquility. There was sufficient food to survive on for the foreseeable future. They could do nothing but wait while Patience Camp drifted slowly northward to break apart.

A bare minimum

By early April 1916 the ice floe had diminished considerably and began to break away around its edges, reducing the area around Patience Camp. The time had come to abandon what was left of it and launch the three lifeboats into the brash – a thick mash of sea ice, icebergs and near-frozen water. There was barely enough room for everyone.

Twenty-eight men and a bare minimum of supplies was all that could travel. This included a bit of food and a sleeping bag for each, the cookers and other absolutely necessary equipment. One unexpected piece of equipment was Leonard Hussey's banjo. Shackleton understood the role Hussey could play in keeping the men's spirits up, thanks to his musical ability.

The lifeboats were loaded until they were nearly awash. The men would sail in them for another 160 miles (257 km) across the open sea to the nearest land, the uninhabited Elephant Island – an island so remote, barren and unforgiving that even whalers had never visited it.

Shackleton and Wild were in command of the *James Caird*. At 23 feet (7 m), it was the largest of the three lifeboats. Frank Worsley, captain of the *Endurance* and third officer Tom Crean, who had distinguished himself on both of Scott's previous expeditions, oversaw the other two.

After six days and nights – by some miraculous combination of fearless leadership and luck in not getting swamped or crushed by the floating icebergs – the three seriously overburdened boats and their dreadfully weakened crews made it across the rough seas to come aground on Elephant Island on April 16.

It was the first solid land their feet had touched since departing South Georgia, fourteen months earlier. According to Frank Wild, "Ten of our party of twenty-eight men were more or less off their heads and eight were rather frost bitten." The men tumbled around the rocky beach in a delirium of joy and relief.

Shackleton ordered their first hot meal in five days. But he soon realized that when a high tide and a strong wind combined forces on their fifteen yard-wide beach (13.7 m), their temporary camp would be under water, and whoever remained on land would be swept away.

Despite their overwhelming relief at reaching land, they would have to move on. They had to seek a safer, higher place to make a permanent camp if any of them were to survive the coming Antarctic winter.

Chapter 40
Survival

Cape Wild

On the day after that landing, six of the bravest and fittest, with Frank Wild in command, climbed back into the *James Caird* lifeboat and set off along the forbidding, ice-choked shore. After nine hours they returned. They had found a better beach and a more secure landing place seven miles (11 km) to the west. But only one.

After another mug-up of good, hot hoosh, the men were coaxed back into the boats to move to the better location Wild and the others had found.

Once all safely ashore, they started to assemble the camp that would be home for most of them for months. It was only another low beach, a fifteen-yard-wide (13.7 m) spit of land between the unclimbable mountains guarding the interior of the island, and a rocky outcropping a few hundred yards out into the water. Shackleton named the place Cape Wild. The men called it Cape Bloody Wild. But there was fresh water from the snow melt, and enough penguins and seals to provide food, and blubber to serve as a cooking fuel.

No one in the world knew of their plight. In those days before instantaneous radio communication, it would not be unusual for a ship to be out of contact for six months or longer.

No outsider would have guessed that the *Endurance* had been crushed and sunk to the bottom of the Weddell Sea, and that the 28 survivors were camped on the shore of an inhospitable, unexplored island far from any seaway. They would never be found there.

Shackleton was determined to take five men in the *James Caird* and sail it another 800 miles (1,287 km) across the roughest seas in the world back to South Georgia, to seek help.

He placed his most trusted and most competent man, Frank Wild, in command of the men to remain on the beach at Cape Wild. Shackleton knew he would be up to the task. Wild had, after all, successfully managed the seven men of his Western Party on Mawson's 1911-1914 Australian Antarctic Expedition. In that case, that small team was 1,500 miles (2,400 km) from help if anything went wrong. Even so, here on Elephant Island, it would not be an easy or simple task to keep the men's spirits up and their health intact, when their very survival was far from assured.

Everyone knew that the *James Caird* and her crew (Shackleton, Worsley, Crean, McCarthy, McNish and Vincent) would be far more likely to founder in the rough seas of the Southern Ocean, taking the remaining men's hope of survival with them. There was not enough food on hand to sustain them through the coming months of waiting. They would have to survive on killing penguins and seals. Their threadbare clothing and sleeping bags would never keep them warm enough. Their tattered tents would soon blow to ribbons. To help with that, they overturned the two remaining boats, making them into a rude hut with barely enough room to house them all.

"Lash up and stow"

Frank Wild was the perfect leader of these desperate 22 men. They knew and respected him, and every man also knew Wild would never betray their trust.

Using bits of crowd psychology he had learned from working with Shackleton during the long months at Ocean and Patience Camps, he gave them enough work – almost, for there was not much to do – to keep them busy and their minds occupied. It imparted enough hope to sustain them during the long weeks of waiting.

Every morning Frank Wild called out, "Lash up and stow, boys,

284 | Dangerous Situations

the Boss may come today." Roll up your sleeping bags and be ready to leave the moment Shackleton arrives, as he certainly will return.

But, the Boss was longer in returning than anyone could have guessed.

At best, sailing to South Georgia should take about three weeks; arranging a rescue vessel should take no more than a week; and the journey to Elephant Island in a larger, faster ship would be one or two weeks.

For four-and-a-half months, the weeks wore on and there was no sign of Shackleton with a rescue vessel. Indeed, everyone began to realize that he might never come. Hope began to dim. But Wild would not let the light go out.

James Hussey wrote of Wild in his diary. "Even when things were looking extremely black, towards the end of our stay on Elephant Island, he was always ready with a joke. It is not too much to claim that we owed our lives to his leadership. He certainly justified the confidence Shackleton had placed in him."

Rescue

The words Frank Wild had steadfastly sung out in the morning to the marooned men on Cape Wild on the day of their rescue were the same words he had used every day before that.

On August 30, 1916, they had not yet "lashed up and stowed" when one of the men outside their miserable hut spotted a ship on the horizon. It was coming closer. In a matter of hours, they were all safely on board the steam tug *Yelcho* and on their way to safety.

It was on the *Yelcho* that Wild and the men of Elephant Island learned about the challenges that Shackleton had endured in rescuing them. Shackleton and the five men on the *James Caird* had survived the open boat journey to South Georgia – the greatest small-boat journey ever undertaken – but the ordeal of the *James Caird's* men had not ended there.

They had landed on the uninhabited side of South Georgia. After a few days of rest, Shackleton, Worsley and Crean – leaving McCarthy, McNish and Vincent in camp to recover – set off on a do-or-die-trying hike across the uncharted, never before climbed, interior mountains of South Georgia. Their goal was to reach the Stromness whaling station on the other side of the island.

After thirty-six hours of hiking with little food and virtually no rest, this very small team – Shackleton, Worsley and Crean – made it. Even today, for trained mountaineers with modern climbing equipment, this is a daunting and challenging climb.

Shackleton immediately arranged for a rescue ship but in the dead of the Antarctic winter, no ship was able to break through the sea ice to reach Elephant Island. Shackleton didn't give up. It was only on the fourth attempt that a ship broke all the way through to reach Cape Wild. By then, it was said that Shackleton's hair had turned grey.

Thanks to Shackleton and Wild, all the men of the *Endurance* survived their ordeal.

One of the survivors named Thomas Orde Lees wrote in his diary, paying tribute to ". . . our splendid, capable leader Wild who by his buoyant optimism, dogged determination, unrivaled experience and calm demeanor had pulled us through these trying months of waiting." These were words that equally applied to Shackleton's efforts.

Most, but not all of the men of the Ross Sea party on the other side of Antarctica survived. Shackleton, Worsley and Crean went on to New Zealand to see about their relief. Frank Wild returned home to England. By the time the men got home in the autumn of 1916, World War I was raging. They, like almost everyone else from the expedition, immediately joined the war effort. Some of the men who had survived the *Endurance* shipwreck, did not survive combat in that war.

A new quest

Shackleton and Wild parted for a while, each to their own roles in that conflict. For a time, the war consumed their boundless energies, but once peace came again, both began to get restless. The two old Antarctic veterans pursued other enterprises for a while, trying to ignore that old siren call of life and adventure amid the deep-frozen wastes.

Shackleton conceived a plan to explore the still uncharted seas of the Arctic. He eventually found a backer for his plan in his old school chum, John Rowett. Rowett was a man of sufficient means and enthusiasm to fully fund Shackleton's next expedition. Wild, still at loose ends, was only too happy to lend a hand, again as second in command.

The plan evolved from seeking unknown lands in the Arctic to exploring the Southern Ocean near Antarctica, in a variety of ways. Shackleton purchased an underpowered, schooner-rigged Norwegian whaler and renamed her *Quest*, to seek out islands long showing on the chart as "doubtful" and prove their existence one way or the other. There were no more Poles to conquer, but the Antarctic region known as Coats Land had yet to be explored and new scientific data acquired. Frank Wild would again be the Boss's second in command.

In a way it was an Antarctic reunion voyage.

Eight of the twenty men in the *Quest's* crew had been with the *Endurance* and, still hankering for a taste of the ice, signed on. The expedition suffered like that one did from insufficient funds and overly optimistic haste. The *Quest* put in for repairs at several ports en route to South Georgia. On the way there before proceeding South, Shackleton suffered a fatal heart attack and died on January 5, 1922. At the request of his widow Emily, he was buried on South Georgia, where his tombstone remains the object of many a pilgrimage by Antarctic enthusiasts everywhere.

In Wild's book *Shackleton's Last Voyage*, he paid heartfelt tribute to his long-time friend. "Here I said goodbye to the Boss, a great explorer, a great leader and a good comrade. No one knew the explorer side of his nature better than I, and many are the tales I could tell of his thoughtfulness and his sacrifices on behalf of others, of which he himself never spoke."

Wild continued the expedition, now as leader, knowing the Boss would have wanted him to. "Having put my hand to the plough, there was no turning back." But without Shackleton's inspirational leadership, the spark had gone out for everyone.

The expedition settled the nonexistence of some previously charted islands and appearances of land at the mouth of the Weddell Sea. An attempt to reach the Antarctic mainland was foiled by the sea ice when the ship was beset in February 7, and carried by the drift for six weeks until March 21. Running low on food and fuel, the *Quest* called at the deserted beach on Cape Wild, on Elephant Island to stock up on elephant seal blubber and penguin meat, then back to the whaling stations on South Georgia for a final visit and farewell.

It was Frank Wild's last Antarctic expedition. The powerful bond that had joined him and Ernest Shackleton in so many ways, that essence of true teamwork, had reached its end. Wild moved to South Africa with his wife, trying his hand at farming and mining. After a divorce, he remarried in 1931, listing his profession on the marriage license as "explorer." He died August 19, 1939, at the age of sixty-five, and was cremated.

Frank Wild's ashes laid in an urn lost in the basement of a South African church until the British Antarctic historian and author, Angie Butler, conducted the detailed detective work to locate the church and find the urn, and arrange a proper burial.

On November 27, 2011, at Grytviken, South Georgia, Frank Wild's ashes were laid to rest alongside his great friend, Sir Ernest Shackleton. The two men were among the greatest of all the very small teams in the history of exploration. Their geographical discoveries included the Beardmore Glacier and a route to the South Pole. In the process, they set an example for endurance and generosity of spirit that can strengthen every extreme of human endeavor.

★

In the final chapters we look at the lessons to be gleaned from all of the very small teams mentioned in this book. What can we take from their experiences to improve our own well-being and chances for success?

Part 4
Lessons

Chapter 41
Lessons

What Can We Learn From All This?

When we started the research for this book, we made lists upon lists of every small team that we could think of. Once we had winnowed those down into a manageable (but still large) single list, we assessed what we had. Some teams we knew a bit about but as we quickly found out, there was much more we did not know, especially when viewed with a focus on teamwork. Other teams we had to research from scratch.

Our work was fun and fascinating. We were sure that by the end of the process it would be easy to discern what made small teams successful. We were surprised to find that every time we thought we hit upon a commonality showing why very small teams worked well together and why other teams faltered, we found another example that disproved our theory.

Here are some examples.

Misconception 1: Team members must get along with each to achieve great results.

Gilbert and Sullivan disproved that theory. They didn't even like being in the same room with each other.

Misconception 2: Team members must respect each other's contributions.

Yet, Admiral Peary after the end of their quest barely even acknowledges Matthew Henson's participation.

Misconception 3: Team members should be friends.
Hillary and Tenzing weren't friends at the outset of their partnership. They were climbing companions within a larger team. Their friendship developed subsequently. Woodward and Bernstein also did not start out as friends.

Misconception 4: Team members must agree on everything.
Susan B. Anthony and Elizabeth Cady Stanton had strong and opposite opinions on whether to advocate for women's rights during the US Civil War. Wilbur and Orville Wright were known to have long and loud disagreements on glider and wing design.

Misconception 5: Team members have to be equals, and
they have to put in equal amounts of work and effort.
As we found, the work was often intrinsically unequal. Some endeavors required specialized talent that only one team member could provide, such as Charles Eames' focus on creating strong, bendable plywood, or George and Ira Gershwin's divergent talents for music and lyrics.

So, what does make for highly successful, very small teams?

More than anything, it takes two specific self-beliefs.
The first is at an individual level, "We are the ones to do this."
The second is at a team level, "We have faith in us as a team that we are best placed to defy the odds and achieve our goal."

These combined beliefs enabled the very small teams in this book to overcome setbacks and obstacles – any of which could have brought the work to a complete halt. Among them were frostbite, scurvy, snow blindness, starvation, injuries, death of companions, strong differences of opinions, death threats, plane crashes, ridicule, oceanic storms, high-altitude death zones, and shipwreck.

These beliefs also helped the teams and the individual team members ignore the naysayers, doubters, and critics – sometimes of the inner variety – who would say:
"This can't be done."

"It's too risky."

"You'll never get it off the ground."

"You won't come back alive."

"It will be panned by the critics."

"It will close after the first performance."

"You'll run out of money/time/patience…"

"Far smarter/better funded/more creative people will do it first." or "You will certainly die if anything goes wrong."

Armed with these two fundamental beliefs – that "we are the ones to do this," and "we have an unshakable faith that our team is best placed to achieve its goal," – each very small team succeeded in their epic endeavor.

Individuals on the team

Not everyone has a collaborative personality. Some people are too self-possessed, self-centered, competitive, or too much of a loner to work well in a very small team. This single fact is essential to appreciate. A team with a self-referential, self-impressed person on it is unlikely to go on to greatness.

From the perspective of individuals on the team, what is needed?

1. Each person needs to have a collaborative personality and a belief that working with others can be its own reward. Also, that cooperation is more important than individual recognition, and a firm belief that their collective synergy will yield a far greater result than each team member could *ever* achieve working on their own.

2. All team members need to understand intrinsically that gamesmanship and internal competitiveness will undermine the team. Holding on to secrets, holding back knowledge or discoveries, gossiping about team members, taking glee in ridiculing others, and bullying can all disintegrate a team quickly.

3. Each individual needs to value the importance of compromise and adaptability. Each team member must be able to accept that no matter how great their own idea is, another team member may have an even better idea, or could point out a flaw that the originator never considered.

4. Individuals cannot be solely motivated by financial reward, power, recognition or fame. Instead, their motivation must come from the rewards of working with others to create something greater than they could have created on their own, and to believe that what they are working on is monumental and important.

The rewards – whether they are money, glory, trophies or fame – will come later, if they come at all. And this must be shared with one's teammates.

5. Trust is an essential part of teamwork. Each team member needs to be both trustworthy as well as trusting of the others on the team. Trust is earned and trust needs to be extended to others, with the understanding that an outsider's talents could very well help the team's core mission. Being part of a very small team should not be so restrictive that individuals cannot branch out and pursue other actions.

The team itself

There are also team dynamics that determine whether a very small team will be successful.

1. Recognize it is a long-term game and not a quick win. Some of the team endeavors described in this book took years of hard work and collaboration to succeed. Susan B. Anthony and Elizabeth Cady Stanton worked together for over 50 years. Peary and Henson explored the Arctic regions together for two decades.

2. Be willing to pursue ideas that are outside the status quo of conventional wisdom. The Wright brothers had to refute Lilienthal's mathematical tables; Woodward and Bernstein identified crimes at the Watergate Hotel that no one initially believed were anything more than a simple burglary.

3. Accept that the team's efforts, no matter how great, can always in the end amount to nothing. Tom Bourdillon and Charles Evans were excellent mountaineers but challenges with the Closed oxygen system prevented them from reaching the top of Everest. Wilson, Bowers and Cherry-Garrard's quest for unhatched emperor penguin eggs could have resulted in the non-retrieval of eggs. As it was, their winter journey only resulted in proving the eggs did *not* link early penguin development to reptilian dinosaurs. The value was not in the prize, but in the camaraderie and shared experiences that came from the team striving for something monumental.

4. Understand that the team needs to be composed of people with complementary, not competing or overlapping skill sets. Each person needs to bring something unique to the team. In addition, the team's hierarchy should always be respected.

The Apollo 13 team was purposely selected to have complementary skill sets. Jim Lovell was the commander and primary decision maker. In Antarctica, Frank Wild recognized Shackleton as "the Boss" and referred to him as such in front of the men on the *Endurance* Expedition. Also, the comedy duos featured in our Hollywood chapter each brought their unique personality and brand of humor to the team.

5. Be willing to accept that while work in a team is not always shared equally, rewards are almost always shared equally. If a team consists of three people all working toward a common goal, the work will not always be split into thirds. The team goal is what's important. To achieve it requires buy-in of the group. Counting who contributed what amount undermines that effort.

In the quest for women's rights, Susan B. Anthony did more in the beginning than Elizabeth Cady Stanton, due to Elizabeth having to care for her young children. To them, it was still an equal partnership.

Additional commonalities of the successful very small teams in this book were:

1. Team members often described themselves using metaphors. They likened themselves to Arctic explorers striving for something elusive and far in the distance, to navigators on ships charting new paths, and to mountain climbers ascending the highest peaks. In other words, they took inspiration from other very small teams like the ones mentioned in this book.

2. Team members rarely if ever panicked, rarely showed distress, and never gave up. This could be due to a sense that the team dynamics required a stiff upper lip and a can-do attitude. More likely, teams gave a sense of security that even if one member was worried, scared or in serious danger, the others were right there with them.

3. Team members always bolstered each other's spirits. In 1902, when Wilbur Wright exclaimed no one in the next thousand years could solve the mystery of manned, powered flight, it was Orville who convinced him that they should go back to their bicycle workshop and revisit their experiments. Their first manned flight occurred the following year.

4. Physical discomfort was often an accepted part of the process. There were many examples of this. Woodward and Bernstein met their source, Deep Throat, very late at night in poorly lit, possibly dangerous parking garages. In the mid to late 1800s, Susan B. Anthony and Elizabeth Cady Stanton traveled alone or together, crisscrossing the United States to attend women's rights conferences. The Wright brothers endured the windswept sands and mosquitoes of Kitty Hawk, North Carolina. The explorers in the North and South Polar regions, the climbers on Everest, and the three men in Apollo 13 in their crippled, unheated spacecraft all endured extreme privations we can only imagine.

5. The teams needed to be highly creative. That was of course true for the teams that featured in the Creative Arts and Hollywood chapters. It was equally true for the other teams. Examples of this were in every very small team. The Wright brothers needed to build two wind tunnels for experimentation of conditions, as well as a lightweight engine. Everest teams needed to build lightweight oxygen systems. Peary and Henson had to build better sledges than the Inuit traditionally used. Woodward and Bernstein needed to develop creative ways to protect Deep Throat's identity.

6. All teams worked well with others. Highly successful, very small teams made non-team members feel valued and included. Gilbert and Sullivan worked closely with D'Oyly Carte. Hillary and Tenzing needed to work well with all the other 1953 British Everest Expedition team members, the Sherpas and other support staff.

Other examples include Josiah and Eleanor Creesy working with all the other officers and crew members on the Flying Cloud clipper ship. The Apollo 13 crew was dependent on Gene Kranz and the rest of the Houston-based NASA team to get them home safely.

★

The last chapter focuses on bringing these lessons into your life, career and endeavors, and reveals the biggest lesson of all.

The Most Important Lesson

Whether or not you think of it this way, you – our readers – are all part of very small teams, and probably more teams than you realize.

You may have noticed your own natural bias toward joining creative forces with another person. Think of your best friend, your spouse, one of your siblings, or a co-worker – someone with whom you must find a solution to a thorny conundrum. Perhaps someone in your club or social group who sees the world through a lens similar to your own.

Given the scope and complexity of today's corporate and government worlds, two or three individuals working together may seem too small to effect any change. Often it is larger teams that exist within organizations to tackle the larger goals. This has evolved into corporate systems rife with committees yet it may be the smaller teams who get the work done.

It is those two or three people combining their energies, intellects and creativity in an intimate manner that changes the casual "what if?" conversations. Through deep research, and often a shared risk, it becomes about who can open the doors of possibility the widest.

The greatest lesson is simply this: to tackle the weightiest, most difficult challenges, a very small team could be the very best configuration.

After all, it took very small teams of just two or three people to invent the airplane; to be the first atop the tallest mountain in the world; to have the greatest influence on modern culture; to create the most compelling adventures in polar exploration; and to deliver the greatest change in women's rights the world has ever seen.

At its heart: Navigation

There were commonalities of endeavor within the teams we profiled. They all brought an elegance to their solution. Many faced adversity, risk and danger. Some overcame issues of prejudice and discrimination on race and gender.

There was another commonality.

They were all navigators. Navigation in the sense of finding the best way through an unmarked, previously unvisited wilderness.

This may be more obvious in the case of Captain and Mrs. Creesy charting a path through the trackless oceans with only a sextant, chronometer, and a yet untried instructional book explaining how to make the best use of the wind, seas and currents. Henson and Peary likewise crossed the shifting ice floes of the Arctic Ocean in dogged pursuit of an unmarked point at the top of the globe. The three men in Apollo 13 had even less to mark their transit through the heavens when all their carefully laid plans went awry.

But think of these ideas as well.

Susan B. Anthony and Elizabeth Cady Stanton forged a way through an even more difficult terrain – the violently shifting concerns of public opinion in the important matter of women's rights. Woodward and Bernstein helped create the process of investigative reporting that now protects people of all nations from the lies and half-truths that find their way into our politics, media and our public discourse. The Wright brothers in their way had a different navigational challenge: to understand the invisible flow of air over physical surfaces, and to understand the push and pull of the four elements of flight (lift, drag, thrust, and weight) that would open the way to heavier-than-air flight.

Gilbert and Sullivan created a whole genre of musical theater that now informs so much of our cultural sensibilities. Ray and Charles

Eames transformed commonplace materials into beautiful and elegant furnishings in ways that had never before been dreamed of, and to this day their designs still look modern.

All of these very small teams worked their miracles by combining the intellects and energies of two, or at most, three people, plunging headlong into yet undiscovered realms of creativity and imagination with an intimacy that can't be achieved by a larger committee or consensus. Their achievements made use of that same energy existing within all of us, enabling us within our own very small teams to make our own world a safer, more productive, more beautiful place.

In conclusion, we urge you to take a few moments to reflect on how all these ideas come into play in your own lives, in your own close work with others as you work together to pursue your joint passions. Not all teams make history or change the world, but some – yours – can bring unimaginable rewards, and maybe even change the course of your own life.

What about *you*? Who's on *your* team?

A note to our readers

We hope you enjoyed this book. We would be most grateful for a review on Amazon, Goodreads or any other sites that readers might visit. Thank you in advance for doing this. It means a lot to us.

This is the third book co-written by our own very small team of Brad Borkan and David Hirzel. Our books draw on the experiences of heroes from the past to teach us how to make better decisions in the future.

We are currently working on several additional books.

We love hearing from readers and can be contacted via our website: *https://www.extreme-decisions.com*. Please join our mailing list. Sign-up details are on our website.

The opening chapter of our first co-authored book, *When Your Life Depends on It: Extreme Decision Making Lessons from the Antarctic*, appears in the Appendix, along with other notes relevant to this book, including a Timeline and a complete list of the many very small teams mentioned in this book.

Brad and Dave

Appendix

Timeline

1480-1490

1488 Leonardo da Vinci designs flying machines in his notebooks.

1780-1790

1783 Joseph and Jacques Montgolfier build the first hot-air balloon and fly a duck, a rooster and a sheep in the basket attached to it.

1800-1810

1804 George Cayley builds the first full-size glider that could be flown like a kite.

1810-1820

1815 Elizabeth Cady Stanton is born.

1820-1830

1820 Susan B. Anthony is born.

1829 George Stephenson and his son Robert invent the first locomotive designed specifically to pull a passenger train.

1840-1850

1840 Elizabeth Cady marries Henry Stanton. Elizabeth insists her marriage vows be rewritten to omit the word 'obey.'

1841 Captain Josiah Creesy marries Eleanor Prentiss.

1848 Seneca Falls Convention for Women's Rights is organized by Elizabeth Cady Stanton and Lucretia Mott.

1850-1860

1850 First National Women's Rights Convention is held in Worcester, Massachusetts.

1851 Eleanor Creesy pilots the brand-new clipper ship *Flying Cloud* to a record-setting maiden voyage, from New York to San Francisco.

1851 Susan B. Anthony and Elizabeth Cady Stanton meet for the first time.

1854 Eleanor and Captain Josiah Creesy in the *Flying Cloud* break their own record for a New York-San Francisco voyage by 13 hours. The record stands until 1989.

1858 Jacques Offenbach's most popular operetta *Orphée aux Enfers* is first performed in Paris. Offenbach's work heavily influences Gilbert and Sullivan.

1860-1870

1860 New York state legislature passes the Married Women Property Act enabling women to manage their own business, property and money.

1861 United States Civil War begins. Susan B. Anthony and Elizabeth Cady Stanton argue over the push for women's rights during the war years.

1865 Mt. Everest is named after the British Surveyor General Sir George Everest, nine years after being confirmed as the tallest mountain in the world.

1869 Wyoming becomes the first US territory or state to grant women the right to vote.

1870-1880

1870 Alphonse Penaud builds a helicopter top that could spin and seemingly "fly."

1870 Gilbert and Sullivan meet for the first time.

1873 Susan B. Anthony is arrested when she tries to vote.

1875 Gilbert and Sullivan's first collaboration, *Trial by Jury,* is first performed.

1879 Gilbert and Sullivan's *Pirates of Penzance* is first performed.

1880-1890

1881 The Savoy Theatre in London opens, purpose-built to stage Gilbert and Sullivan operettas. It is the first public building in the world to be lit entirely by electricity.

1885 Gilbert and Sullivan's *The Mikado* is staged for the first time.

1886 Peary's first attempts to cross the north Greenland icecap.

1887 Peary meets Henson for the first time and hires him.

1887 Arthur Conan Doyle creates the crime solving duo, Sherlock Holmes and Dr. Watson.

1890-1900

1890 Gilbert and Sullivan break up in a dispute over the cost of a carpet.

1891 The start of Peary's second Greenland expedition, with his wife Josephine and his new exploring companions, Matthew Henson and Eivind Astrup.

1892 Charles Rennie Mackintosh and Margaret Macdonald meet for the first time at the Glasgow School of Art.

1892 Elizabeth Cady Stanton, at the age of 72, writes the *Solitude of Self* speech and delivers it a the National American Woman Suffrage Association convention and to the US Senate judiciary committee.

1892 The Wright brothers open their bicycle shop in Dayton, Ohio.

1892 Peary makes his second attempt to cross the Greenland icecap.

1895 First human sets foot in Antarctica.

1896 Lilienthal dies in a glider accident after 2000 successful flights, inspiring the Wright brothers to get more interested in flight.

1898 The 50th anniversary of the first women's rights convention in Seneca Falls, New York. Elizabeth Cady Stanton is too old to attend.

1898 The start of Peary's expedition to explore north Ellesmere Island.

1899 The Wright brothers begin their efforts to understand flight.

1900-1910

1900 Wright brothers go to Kitty Hawk, North Carolina, for the first time to experiment with their glider.

1901 Shackleton and Wild meet during Scott's *Discovery* Expedition.

1902 After significant setbacks, Wilbur exclaims that he doesn't think a flying machine could be built in the next one thousand years.

1903 Wright brothers achieve the first powered flight with the airplane they call the *Flyer*.

1906 Susan B. Anthony dies. Four US states have granted the women the right to vote by the time of her death.

1908 Shackleton and Wild reach their "furthest South," 97 miles (156 km) from the South Pole before having to turn back.

1909 Peary claims (falsely) that he has reached the North Pole.

1909 Wright brothers establish the Wright Company which builds airplanes and runs a flying school.

1910-1920

1910-1913 Scott's *Terra Nova* Expedition.

1911 Wilson, Bowers and Cherry-Garrard undertake their quest for penguin eggs in the Antarctic winter, commemorated in Cherry-Garrard's book, *The Worst Journey in the World*.

1912 Scott, Wilson, Bowers, Oates and Evans reach the South Pole, and then perish on the return journey.

1912 Wilbur Wright dies.

1914-1916 Shackleton's Imperial Trans-Antarctic Expedition takes place. Also called the *Endurance* Expedition.

1916 Alfred Stieglitz and Georgia O'Keeffe meet for the first time.

1916 Frank Wild instructs the 21 other men marooned on Elephant Island to "Lash up and stow. The Boss may come today." Four and a half months later, Shackleton does rescue them.

1919 The 19th amendment of the United States Constitution granting American women the long-sought-after right to vote passes Congress in 1919. It is ratified in 1920.

1920-1930

1921 The First British Everest Reconnaissance Expedition is launched from the Tibet side.

1922 Ernest Shackleton dies aboard the *Quest*. He is buried at South Georgia.

1922 Seven Sherpas die in an avalanche during the Second British Everest Expedition.

1924 Alfred Stieglitz and Georgia O'Keeffe get married.

1924 George Mallory is asked why he wants to climb Everest. He replies, "Because it's there."

1924 Edward Norton, climbing without oxygen, achieves a high-altitude record on Everest that stands for 28 years: 28,100 feet (8,560 m).

1924 Mallory and Irvine die on Mt. Everest. We will never know if they got to the top of the mountain.

1925 Oscar Hammerstein II starts partnering with Jerome Kern. They go on to create the musical *Show Boat*.

1927 Charles Lindbergh flies solo across the Atlantic. Orville Wright meets him in Paris.

1930-1940

1933 First flights occur over Everest.

1935 George and Ira Gershwin's *Porgy and Bess* first opens on Broadway.

1935 Tenzing Norgay is a Sherpa for the first time in Everest, helping the Fifth British Everest Expedition.

1938 Abbott and Costello's *"Who's On First?"* is first broadcast.

1939 Frank Wild dies in South Africa. His ashes are later buried beside Ernest Shackleton's grave in South Georgia, in 2011.

1940-1950

1940 Charles and Ray Eames meet at the Cranbrook Academy of Arts for the first time.

1940 Robin is first introduced in a Batman comic book.

1941 Morecambe and Wise first partner together and start their long running British comedy act.

1943 *Oklahoma!*, written by Rodgers and Hammerstein, opens for the first time on Broadway. It runs for over five years on Broadway, breaking all previous records.

1948 Orville Wright dies.

1949 Nepal opens its borders to Everest climbers.

1950-1960

1951 The British Reconnaissance expedition to Everest is led by Eric Shipton. His team includes Tom Bourdillon, Ed Hillary, Michael Ward and others.

1952 On the Swiss Everest Expedition, Raymond Lambert and Tenzing Norgay ascend to a new all-time altitude record: 28,210 feet (8,600 m).

1953 Hillary and Tenzing achieve the first summit of Mt. Everest. It is announced to the world on same day as Queen Elizabeth's coronation.

1953 The state of Oklahoma adopts the theme song from Rodgers and Hammerstein's *Oklahoma!* as its state anthem.

1955 Matthew Henson dies.

1956 Charles and Ray Eames design the iconic Eames chair.

1958 Christo and Jeanne-Claude meet for the first time.

1959 *Rocky and Bullwinkle* cartoon first airs.

1960-1970

1960 *The Flintstones* cartoon first airs featuring Fred and Barney, and their wives Wilma and Betty.

1961 Apollo space program begins.

1964 Charles and Ray Eames' *Powers of Ten* video is shown in the IBM pavilion of the New York World's Fair.

1966 *Star Trek* first airs featuring Captain Kirk and Mr. Spock.

1967 Apollo 1's fatal fire kills all three astronauts in the Command Module.

1967 *Bonnie and Clyde* movie is released.

1968 Richard Nixon is elected as the 37th President of the United States.

1969 Neil Armstrong and Buzz Aldrin land on the moon.

1969 Wally Herbert becomes the first man to reach the North Pole using the methods pioneered by Peary and Henson.

1969 Princeton University admits women for the first time, over 100 years after Elizabeth Cady Stanton urged universities to admit women.

1969 Bert and Ernie first appear on Sesame Street.

1970-1980

1970 Apollo 13 experiences problems in outer space on its way to the moon.

1970 Simon and Garfunkel break up.

1970 *The Odd Couple* first airs – a comedy series featuring the mismatched men, Felix and Unger.

1972 Bob Woodward and Carl Bernstein are young reporters assigned to desks in the same row of the *Washington Post* newsroom.

1972-1974 The Watergate hearings begin.

1974 Nixon resigns as a result of Watergate, a case brought to the public's attention by Woodward and Bernstein.

1975 Penn and Teller first perform together.

1975 *Starsky and Hutch* airs on television for the first time.

1976 Woodward and Bernstein publish their book *All the President's Men*.

1976 Steve Jobs and Steve Wozniak build Apple Computer 1.

1977 The first *Star Wars* movie launches featuring the multi-galaxy pair, Han Solo and Chewbacca.

1978 Charles Eames dies.

1980-1990

1987 Matthew Henson's remains are moved to Arlington National Cemetery and buried next to Peary's.

1988 Ray Eames dies, exactly 10 years to the day after her husband Charles Eames died.

1989 Captain Josiah and Eleanor Creesy's 1854 record time in sailing a ship from New York to San Francisco via Cape Horn is finally broken by a high-performance racing sloop. The record stood for 135 years.

1990-2000

1991 *Thelma and Louise* launches in cinemas.

1995 *Toy Story* launches featuring Woody and Buzz Lightyear

1995 *Apollo 13* movie premieres

1996 The worst year for mountaineer deaths on Everest. 3% of the climbers who ascended above Base Camp died ascending or descending the mountain.

1998 *Time Magazine* places Rodgers and Hammerstein in their list of the top 20 most influential artists of the 20ᵗʰ century.

1999 George Mallory's body is found on Everest. Irvine's body is not. No one knows if they had died en route to the summit or descending from it.

2000-2010

2005 Christo and Jeanne-Claude's *The Gates* exhibition appears in New York City's Central Park.

2010-2020

2011 Frank Wild's ashes are laid to rest at Grytviken, South Georgia, alongside his great friend, Sir Ernest Shackleton.

2012 The National History Museum, London opens its Treasures Room showcasing the 22 most important objects from its collection of 70 million specimens. One of the objects is one of the three emperor penguin eggs brought back by Wilson, Bowers and Cherry-Garrard.

2014 The Glasgow School of Art, designed by Mackintosh (the amount Macdonald contributed to the design is unknown), suffers a disastrous fire. It is rebuilt.

2018 The Glasgow School of Art suffers another disastrous fire. It is being rebuilt.

2020-Today

2020 Christo dies. The French government gives permission for the wrapping of the Arc de Triomphe in Paris in silver fabric, a project Christo and Jeanne Claude conceived in the early years of their relationship.

2023 Woodward and Bernstein are popular guests on TV news programs comparing the criminality of the Nixon and Trump presidencies.

2024 NASA delays the launches of Artemis II and Artemis III.

2024 Rodgers and Hammerstein's *Oklahoma!* is on stage currently in London and on Broadway.

2024 Gilbert and Sullivan operettas like *HMS Pinafore* and *The Mikado* continue to be performed by professional and amateur theatre companies around the world.

2025 NASA will launch the Artemis II mission to orbit the moon. Two of the four astronauts onboard will include the first woman and first Black man to reach, but not land on, the moon.

2026 NASA will launch Artemis III. It will be the first mission to land people on the moon since 1972. The four-person crew has not yet been announced, but NASA has committed that at least one of the two crew members who walk on the moon will be a woman.

2030 and beyond. People will continue to marvel at the incredible accomplishments of the very small teams mentioned in this book, and Gilbert and Sullivan operettas will continue to be performed worldwide.

Very Small Teams Mentioned In This Book

Abbott and Costello (Bud and Lou) – comedians

Anthony and Stanton (Susan and Elizabeth) – women's rights activists, suffragettes

Armstrong and Aldrin (Neil and Buzz)– astronauts

Astaire and Rogers (Fred and Ginger)– dancers and actors

Ball and Arnaz (Lucille and Desi)– actors and producers

Barker and Corbett (The Two Ronnies) – comedians

Batman and Robin – fictional crime fighters

Ben and Jerry – ice cream manufacturers

Bert and Ernie – *Sesame Street* characters

Bonnie and Clyde – bank robbers

Boris and Natasha – cartoon characters

Bourdillon and Evans (Tom and Charles) – mountaineers

Boyce and Hart (Tommy and Bobby) – singers

Brunel (Marc and Isambard) – father and son, engineers

Butch Cassidy and the Sundance Kid – train robbers

Christo and Jeanne-Claude – artists

Collins and Lapierre (Larry and Dominique) – authors

Creesy (Josiah and Eleanor) – husband and wife, sea captain and navigator

Pilâtre de Rozier and Laurent (Jean-Francois and Francois) – hot air balloonists

Eames (Charles and Ray) – husband and wife, designers

Felix and Oscar – *Odd Couple* characters

Finger and Kane (Bill and Bob) – authors, illustrators, creators of *Batman*

Flintstone and Rubble (Fred and Barney) – cartoon characters

French and Saunders (Dawn and Jennifer) – comedians

Gates and Allen (Bill and Paul) – software engineers, founded Microsoft

Gershwin (George and Ira) – brothers, composers

Gilbert and Sullivan (W.S. and Arthur) – composers

Hall and Oates – singers

Hammerstein and Kern (Oscar and Jerome) – composers

Han Solo and Chewbacca – *Star Wars* characters
Hewlett and Packard (Bill and David) – engineers, founded HP
Hillary and Lowe (Ed and George)– mountaineers
Hillary and Norgay (Ed and Tenzing)– mountaineers
Holland, Dozier, Holland (Brian, Lamont and Eddie) – songwriters
Holmes and Watson (Sherlock and John) – detectives
Jagger and Richards (Mick and Keith) – songwriters and musicians
Jobs and Wozniak (Steve and Steve) – inventors, founded Apple
Jones and Ravenwood (Indiana and Marion) – *Raiders of the Lost Ark* characters
Kariko and Weissman (Katalin and Drew) – medical researchers, Nobel prize winners
King and Goffin (Carole and Gerry) – singers
Kirk and Spock (Jim and Spock) – *Star Trek* characters
Lambert and Norgay (Raymond and Tenzing) – mountaineers
Laurel and Hardy (Stan and Oliver) – comedians
Lennon and McCartney (John and Paul) – songwriters and musicians
Lovell, Haise and Swigert (Jim, Fred and Jack) – astronauts, Apollo 13
Luhrs, Bergstrom and Hazelton (Warren, Lars and Courtney) – sailors
Mackintosh and Macdonald (Charles and Margaret) – architects, artists, designers
Mallory and Irvine (George and Sandy) – mountaineers
Mann and Weill (Barry and Cynthia) – singers
Montgolfier (Joseph and Jacques) – brothers, inventors
Moore and Noyce (Gordon and Bob) – inventors, founded Intel
Morecambe and Wise (Eric and Ernie) – comedians
Norton and Somervell (Edward and Theodore) – mountaineers

Page and Brin (Larry and Sergey) – software engineers, founded Google

Peary and Astrup (Robert and Eivind) – explorers

Peary and Henson (Robert and Matthew) – explorers

Penn and Teller – magicians

Procter and Gamble (William and James) – manufacturers

Riggs and Murtaugh (Martin and Roger) – fictional cops

Rizzoli and Isles (Jane and Maura) – fictional crime fighters

Rocky and Bullwinkle – cartoon characters

Rodgers and Hammerstein (Richard and Oscar) – composers

Rodgers and Hart (Richard and Lorenz) – composers

Sam and Dave – singers

Scott and Ponting (Robert and Herbert) – explorers and photographers

Shackleton and Wild (Ernest and Frank) – explorers

Shackleton, Worsley and Crean (Ernest, Frank and Tom) – explorers

Shipton and Smythe (Eric and Frank) – mountaineers

Siegfried and Roy – animal trainers, performers

Simon and Garfunkel (Paul and Art)– songwriters and singers

Solo and Kuryakin (Napoleon and Illya) – fictional secret agents

Sonny and Cher – singers

Starsky and Hutch – crime fighters

Stieglitz and O'Keeffe (Alfred and Georgia) – photographer and artist

Stephenson (George and Robert) – father and son, engineers

The Three Stooges (Moe Howard, Shemp Howard and Larry Fine) – comedians

Thelma and Louise – film characters

Todd and Pitts (Thelma and Zasu) – comedians

Tom and Jerry – cartoon characters

Torvill and Dean (Jayne and Christopher) – Olympic champion ice dancers

Turners (Ike and Tina) – husband and wife, singers

Vaughn and McCallum (Robert and David) – actors
Williams (Serena and Venus) – sisters, tennis players
Wilma and Betty – cartoon characters
Wilson, Bowers and Cherry-Garrard (Edward, Birdie and Apsley) – explorers
Wiseman, Glover, Koch, and Hansen (Reid, Victor, Christina and Jeremy) – astronauts
Woodward and Bernstein (Bob and Carl) – journalists
Woody and Buzz Lightyear – *Toy Story* characters
Wright (Wilbur and Orville) – brothers, inventors

Recommended Reading

Have we whetted your appetite to learn more about the teams that appear in this book? Below is a list of books that we have enjoyed and would recommend.

A star (★) has been added to highlight our top recommendations.

Apollo Thirteen

- *Lost Moon: The Perilous Voyage of Apollo 13*, Jim Lovell and Jeffrey Kluger (Houghton Mifflin, 2006)

Eames

★ *Charles and Ray Eames: Pioneers of Mid-Century Modernism*, Gloria Koenig, (Taschen, 2005)

Eleanor and Josiah Creesy

- *Flying Cloud: The True Story of America's Most Famous Clipper Ship and the Woman Who Guided Her*, David W. Shaw, (Harper Collins, 2001)

Everest

★ *Everest 1953: The Epic Story of the First Ascent*, Mick Conefrey, (Oneworld, 2012)
- *Seven Steps from Snowdon to Everest: A Hill Walker's Journey to the Top of the World*, Mark Horrell, (Mountain Footsteps Press, 2015)

Peary and Henson

- *How Peary Reached the Pole: The Personal Story of His Assistant*, Donald B. MacMillan, (Houghton Mifflin, 1934)
- *Matthew A. Henson's Historic Arctic Journey,* Matthew A. Henson with introduction by Deirdre Stam, (Explorers Club Classics, 2009)
- *Peary's Arctic Quest: Untold Stories from Robert E. Peary's North Pole Expeditions*, Susan Kaplan and Genevieve LeMoine, (Down East Books, 2019)

★ *The Noose of Laurels: Robert E. Peary and the Race to the North Pole,* Wally Herbert, (Atheneum, 1989)
• *The North Pole,* Robert E. Peary, (Hodder and Stoughton, 1910)

Scott and Ponting
• *Herbert Ponting: Scott's Antarctic Photographer and Pioneer Filmmaker,* Anne Strathie, (History Press, 2021)
• *The Lost Photographs of Captain Scott,* Dr. David Wilson, (Little Brown, 2011)
• *With Scott to the Pole: Terra Nova Expedition 1910-1913; The Photographs of Herbert Ponting,* Herbert Ponting, Ranulph Fiennes (foreword), (Ted Smart, 2004)

Shackleton and Wild
• *Frank Wild,* Leif Mills (Caedmon of Whitby, 1999)
• *Shackleton,* Margery and James Fisher (Barrie, 1957)
★ *Shackleton: By Endurance We Conquer,* Michael Smith (Oneworld, 2015)
• *Shackleton's Last Voyage: The Story of the Quest. From the Official Journal and Private Diary kept by Dr. A.H. Macklin,* Frank Wild (Cassell, 1923)
• *South: The Endurance Expedition,* Ernest Shackleton (Penguin, 2013) *originally published in 1919*
• *The Heart of the Antarctic,* Ernest Shackleton, (Wordsworth, 2007) *originally published in 1909*
• *The Quest for Frank Wild,* Angie Butler, (Jackleberry Press, 2011)

Wilson, Bowers and Cherry Garrard
★ *The Worst Journey in the World,* Apsley Cherry-Garrard (Penguin, 1937) *originally published in 1922*
• *When Your Life Depends on It: Extreme Decision Making Lessons from the Antarctic,* Brad Borkan and David Hirzel, (Terra Nova Press, 2017)

Women's Rights

★ *Elizabeth Cady Stanton and Susan B. Anthony: A Friendship that Changed the World,* Penny Colman, (Henry Holt and Company, 2011)

Woodward and Bernstein

★ *All the President's Men: The Most Devastating Political Detective Story Of The 20th Century,* Bob Woodward and Carl Bernstein, (Simon & Schuster, 1974)

Wright Brothers

• *How We Invented the Airplane: An Illustrated History,* Orville Wright, edited by Fred C Kelly (Dover Publications, 1953)
★ *The Wright Brothers: The Dramatic Story Behind the Legend,* David McCullough, (Simon & Schuster, 2015)

A note about our research methods

We approach the research for all our books the old-fashioned way: by reading the leading publications on the subject, visiting museums and speaking with experts. All the statements in this book have been checked and re-checked for historical accuracy, using multiple different sources to assess the validity of the statement.

We invited historians to review drafts of the book. When we needed certain elements that could be gleaned from the internet, we relied on trusted sites like the Encyclopedia Britannica, National Geographic, National Park Service, Royal Geographical Society, Scott Polar Research Institute, and the Smithsonian, as well as selected specialist sites. We double and triple checked any information found on these sites with other sources. Specialist websites were used for currency, distance and temperature conversions.

Acknowledgements

All of the very small teams mentioned in this book were assisted and encouraged by a variety of people. To create this book, we have been equally supported by wonderful individuals and other small teams – historians, mountain climbers, polar explorers, museum curators and many other experts.

We'd first like to thank the multi-national creative team we assembled for our previous books and who we used for this one. This included our superb editor Kate Gallagher of Nerd Girl Edits, our book designer Anne Sharples who created the covers and interior of our books, and our social media advisor, Nathan James of Softwood Books.

As part of the research for the book, we visited many locations including the Edward Wilson Gallery in Cheltenham, England; Georgia O'Keeffe Museum in Santa Fe, New Mexico; Kennedy Space Center in Florida; National Air and Space Museum in Washington, D.C.; Natural History Museum in London; Peary-MacMillan Arctic Museum in Brunswick, Maine; Royal Geographical Society in London; Science Museum in London; Scott Polar Research Institute in Cambridge, England; Shackleton Museum in Athy, Ireland; South Georgia (Antarctica) Museum; and the Victoria and Albert Museum in London. We learned from their many displays as well as the curators, archivists and others who work at these wonderful places.

The team at the Royal Geographical Society: Christine James; Eugene Rae, librarian and Everest expert; Jamie Owen, archivist; and Joy Wheeler, assistant librarian, were especially helpful. They enabled Brad to view the original source material from numerous British Everest expeditions. This included the handwritten diary of John Hunt, the 1953 British Everest Expedition leader, where he listed the climbing order for the ascent teams.

There were so many people we talked with and who advised us that the following can only be a partial list, arranged in first name alphabetical order. Angie Butler, Antarctic historian, author and Polar travel agent, biographer of Frank Wild, and the finder of Frank Wild's ashes in South Africa. Alexandra Shackleton, granddaughter of Sir Ernest Shackleton, shared many stories of her grandfather. Anne Strathie, an Antarctic historian, author, and biographer of Birdie Bowers and the photographer Herbert Ponting. Bob Headland, the world's leading authority on polar exploration, provided valuable advice. Barbara Rae, polar artist and member of the Royal Academy of Art, shared her experiences of travel in the Arctic as well as some photographs, one of which appears in this book.

David Wilson, Antarctic historian, author, great-nephew of Dr. Edward Wilson, and expert on the Scott and Ponting photographic team. Deirdre Hanna, who as a young girl was at Queen Elizabeth's Coronation in 1953, described for us the excitement in the crowd when news of the successful Everest climb by Hillary and Tenzing was announced. Deirdre Stam, expert on Peary and Henson. Douglas Russell, curator at the Natural History Museum provided valuable insights into Wilson, Bowers and Cherry-Garrard's quest for the penguin eggs, as well as on how the eggs were preserved.

Helen Wernham shared her experiences of visiting Ross Island, Antarctica and provided the photo of the Emperor penguin. Jan Chojecki, Antarctic historian, author and grandson of John Quiller Rowett who funded Shackleton and Wild's last expedition. John Kafalas provided useful corrections on the musical theatre chapters. Kelsey Rose Williams, the Eames Office archivist, provided important edits to the Eames section as well as provided the Eames photographs. Mark Horrell, mountaineer and author who climbed Everest, provided insights into the challenges of the climb. Michael and Angela Galper provided photographs of the Watergate complex. Michael Smith, polar historian, author and biographer of Shackleton was a go-to source for numerous insights on a variety of the teams in this book.

Seb Coulthard, historian, polar adventurer and lecturer, shared his deep knowledge with us. Shari Powell provided useful material about Kitty Hawk, N.C. and the Wright brothers. Stephen Scott-Fawcett, historian and Everest climber who manages the Shackleton Appreciation facebook group has always been a keen supporter of our writing endeavors. Stuart Leggatt shared numerous insights that were helpful in the chapters on mountaineering and polar explorers.

We would like to thank the many people who also encouraged the writing of this book: Ann Richardson, Bonnie Graham, Don Fishbein, Hazel Stix, Helen Dwight, Holly Worton, Karen Carpenter, Leena H., Nicola Rossi, Shari Powell, Stella Paris, Sue Bennett, and Sue Quelch.

For his Antarctic trip, Brad would like to express his deep-felt thanks to Angie Butler and Caro Mantella of the Ice Tracks Travel Agency and to the crew and staff of the *Island Sky* ship run by Polar Latitudes, as well as to his seventy-seven shipmates on the Final Quest voyage in January 2023 retracing some of Shackleton and Wild's steps.

Our website *www.extreme-decisions.com* was created by Leah Matthews of Virtual Studio and supported by Jacques Soudan.

Thanks also to our families and close friends who, for the past three years, listened patiently while we regaled them (probably too many times) with stories of the amazing power of very small teams in the quest for women's rights, building the first airplane and why only the Wright brothers could have done it, how Gilbert and Sullivan didn't like each other, the astonishing story of the Creesys, and the quests for Everest, the moon and the North and South Poles.

The authors are most deeply indebted to our families for their unflagging support and encouragement. Brad to his wife, Anne and their daughter Brittany, and David to Alice Cochran, the light of his life.

While we gained valuable insights from all the organizations and people mentioned, any errors are our own.

Sources For The Diagrams, Maps, Quotations And Photographs

Below is a complete reference to sources for the diagrams, maps, quotations and photographs that appear in this book. We are most grateful to the Royal Geographical Society, NASA and d-maps.com for permission to use their materials.

Great effort has been taken in selecting photographs that are in the public domain because they are out of copyright or free of copyright restrictions. Details about each photograph are shown in the listing below. All photographs have been sourced from the following locations: Library of Congress, NASA, and the Royal Geographical Society, and other sources indicated. Some of the photographs were taken by friends of the authors who have granted permission to include their photographs in this book.

To keep footnotes to a minimum, wherever a quote appears within a chapter, we indicate the source in the text. Additional details are provided below.

CHAPTER 1: Photograph source: Image of the Women's Declaration of Rights Image from Library of Congress, Manuscript Division, Susan B. Anthony Papers, 1876.
Elizabeth Cady Stanton and Susan B. Anthony [Between 1880 and 1902] Photograph retrieved from the Library of Congress <www.loc.gov/item/97500087/>.
Cartoon: Artist Charles Lewis Bartholomew, [Between 1892 and 1896] Photograph retrieved from the Library of Congress, <www.loc.gov/item/2016678242/>.
Opening quotation source: "The days when the struggle was the hardest..." *Life and Work of Susan B. Anthony*, page 1263.
CHAPTER 2: Elizabeth Cady Stanton's address to the Legislature of New York, 1854: Library of Congress
CHAPTER 4: Elizabeth Cady Stanton's *Solitude of Self* speech: Library of Congress.
CHAPTER 5: Photograph source: Orville Wright, age 34, (photographer: Wilbur Wright) Photograph retrieved from the Library of Congress,

<www.loc.gov/item/2001696608/>.

Wilbur Wright, age 38. One of the earliest published photographs of him. (photographer: Orville Wright) Photograph retrieved from the Library of Congress, <www.loc.gov/item/2001696613/>.

Opening quotation source: "...nearly everything that was done in our lives..." *25 Best Business Partner Duos of All Time.*

CHAPTER 6: Photograph source: Wilbur Wright lying prone in glider just after landing. (Photographer: Orville Wright) Photograph retrieved from the Library of Congress, <www.loc.gov/item/2001696465/>

CHAPTER 7: "Not in a thousand years..." There are numerous sources for Wilbur Wright's quote including the University of Chicago library.

Photograph source: First flight. (Photographer John T. Daniels) Photograph retrieved from the Library of Congress, <www.loc.gov/item/00652085/>.

CHAPTERS 8-10: Photograph source: Peary Photograph retrieved from the Library of Congress, <www.loc.gov/item/00650165/>

Henson: Photograph retrieved from the Library of Congress, <www.loc.gov/item/00650163/>

Arctic ice sheets: Barbara Rae kindly provided this photograph to the authors.

Opening quotation source: "It'll work if God, wind, leads, ice, snow..." *Profile: African-American North Pole Explorer Matthew Henson,* National Geographic, January 2003. Other quotes within these chapters are from MacMillan's book, *How Peary Reached the Pole: The Personal Story of His Assistant*, Henson's autobiography, and Peary's *North Pole* book.

Map source: Arctic outline map from *https://d-maps.com/carte. php?num_car=3189&lang=en.* Annotations added by the authors.

CHAPTERS 11-13: Photograph Source: Bowers, Wilson and Cherry-Garrard. Photographer, Herbert Ponting, 1911. Photograph from *The Great White South* by Herbert Ponting, 1923, facing page 155.

Emperor Penguin. Helen Wernham kindly provided this photograph to the authors.

Upon their return: Wilson, Bowers and Cherry-Garrard. Photographer, Herbert Ponting, 1911. Photograph from from *Scott's Last Expedition,* by Dr. Edward Wilson and the Surviving Members of the Expedition, 1913, Vol. II, facing p. 72. Photograph kindly provided by the Royal Geographical Society. © Royal Geographical Society (with IBG).

Penguin egg: Photographer: Brad Borkan, 2024.

Opening quotation source: "We did not forget the Please and Thank you..." and other quotes within these chapters are from Cherry-Garrard's book, *The Worst Journey in the World*.

Map source: The authors created this map based on Edward Wilson's original map as it appeared in Cherry-Garrard's book. Antarctica outline map from *https://d-maps.com/pays.php?num_pay=292&lang=en*. Annotations added by the authors.

CHAPTER 14-16: Photograph source: Eames House: Photographer: Carol M. Highsmith. This photograph is from the Carol M. Highsmith Archive collection at the Library of Congress. Highsmith has stipulated that her photographs are in the public domain.

Washstand: Public domain photograph. "This file was donated to Wikimedia Commons as part of a project by the Metropolitan Museum of Art."

Scott, Wilson, Bowers, Oates and Evans at the South Pole. Photograph retrieved from the Library of Congress, <www.loc.gov/item/2009633365/> Photograph in *The Great White South* by Herbert Ponting, 1923, facing page 280.

Orange-cloth gates: Photographer: Brad Borkan

Opening quotation source: "Genius? Nothing – we just worked harder." Koenig's book, *Charles and Ray Eames: Pioneers of Mid-Century Modernism*.

CHAPTER 17: Photograph source: Poster of Gilbert and Sullivan operattas: Lithograph from H.A. Thomas Lithograph Studio. Photograph retrieved from the Library of Congress, <www.loc.gov/item/2014635751/>.

Sullivan: This work is in the public domain in its country of origin and other countries and areas where the copyright term is the author's life plus 100 years or fewer.

Gilbert: Image from a cabinet card of W.S. Gilbert printed around 1880 by Elliott & Fry. This work is in the public domain in its country of origin and other countries and areas where the copyright term is the author's life plus 100 years or fewer.

Opening quotation source: "A Gilbert is of no use without a Sullivan." *Gilbert and Sullivan: A Dual Biography.*

CHAPTER 19: Quote about Rodgers and Hart from an article about them in the *Atlantic Magazine,* April 2013.

CHAPTERS 20-22: Photograph source: White House: Photographer: Brad Borkan.

Watergate Hotel: Angela and Michael Galper kindly provided this photograph to the authors.

Opening quotation source: "I hereby resign the Office of President of the United States." United States National Archives. All other quotes within these chapters are from Woodward and Bernstein's book, *All the President's Men.*

CHAPTER 23: Photograph Source: The *Flying Cloud:* Image is from oil-on-canvas painting by Antonio Jacobsen, 1917. "This image is in the public domain in its country of origin and other countries and areas where the copyright term is the author's life plus 100 years or fewer."

Opening quotation source: "A sailing vessel is alive…" *Navigatecontent. com: Epic and Inspirational Sailing Quotes.*

Map source: Western hemisphere outline map from *https://d-maps. com/carte.php?num_car=1417&lang=en.* Annotations added by the authors.

CHAPTER 26: Photograph source: Hollywood sign: Photographer: Carol M. Highsmith. Photo is from the Jon B. Lovelace Collection of California Photographs in Carol M. Highsmith's America Project, Library of Congress, Prints and Photographs Division. Photograph retrieved from the Library of Congress, <www.loc.gov/ item/2013633245/>.

Opening quotation source: "Louise, no matter what happens…" *Thelma and Louise movie.*

CHAPTER 29: Photographic Source: Hillary and Tenzing: Photographer: Alfred Gregory, May 28, 1953. Photograph kindly provided by the Royal Geographical Society. © Royal Geographical Society (with IBG)

Opening quotation source: "There is something about building up a comradeship…" *Brainyquotes.com.*

CHAPTERS 30-33: Quotes from John Hunt's diaries at the Royal Geographical Society, Deirdre Hanna's first-hand account of the announcement of the ascent of Everest at the Queen's Coronation, and Conefrey's book, *Everest 1953.*

Map source: The authors superimposed the locations of the base and advance camps, and the climbing path, onto a photograph taken by Brad Borkan of the Everest model at the Royal Geographical Society. The RGS kindly granted permission for us to include this image in our book. Annotations added by the authors.

CHAPTERS 34-36: Photograph Source: Apollo 13 Spacecraft on launchpad: Photograph NASA Image and Video Library

Mission Control: Photograph NASA Image and Video Library

Apollo 13 crew exiting rescue helicopter: Photograph NASA Image and Video Library

Opening quotation source: "I have never lost an American in space…" *PBS article about Apollo 13 quotes.* The quote cited is Gene Kranz' actual quote. The *Apollo 13* movie was not accurate. Quote from President Kennedy's speech: JFK Library. Astronaut quotes from NASA transcripts.

Diagram source: Drawings of the Apollo 13 spacecraft and its flight path are based on NASA diagrams. NASA's chief historian kindly granted us permission to use these diagrams. Annotations added by the authors.

Chapters 37-40: Photograph Source: Wild and Shackleton: Photographer: Frank Hurley. Photograph kindly provided by the Royal Geographical Society. © Royal Geographical Society (with IBG)

Opening quotation source: "He is my second self…" *Royal Scottish Geographical Society article: Frank Wild: The Call of the 'White Unknown.'* Quotes within these chapters are from Mill's book, *Frank Wild.*

Map source: Antarctica outline map from *https://d-maps.com/pays. php?num_pay=292&lang=en.* Annotations added by the authors.

About The Authors

Brad Borkan has had an interest in how people survive and thrive in almost impossible situations for as long as he can remember. He is an author and lecturer and has presented at business and Antarctic conferences, and has been a guest on numerous business and history podcasts. Prior to becoming a full-time author, Brad was a senior director at leading tech companies like SAP. His talks focus on leadership, teamwork and winning against the odds.

Brad has traveled to all seven continents. Originally from the US and now based in London, Brad has a graduate degree in Decision Sciences from the University of Pennsylvania, and is a Fellow of the Royal Geographical Society. He is also a member of the Society of Authors, vice chair of the Friends of the Scott Polar Research Institute, and has given guest lectures on two Antarctic cruise ships.

Brad is also the co-author of *Dynamic Alignment: The Power of Finding Your Purpose, Achieving Your Goals, and Living a Passion-Driven Life.*

David Hirzel is a maritime historian, author and small business owner. His enduring interest in world history and the evolution of societies focuses on the study of polar exploration, and his vocation in architecture expands those interests.

David works out of his design office overlooking the sea in Pacifica, California, with a drafting table, a wide desk and an even wider library of technical and history books to hand. His polar books include a three-part polar biography of the Irish explorer Tom Crean, who was a key player in Captain Scott and Ernest Shackleton's expeditions. David is also a popular lecturer on board polar and maritime exploration cruise ships.

Opening Chapter from *When Your Life Depends on It: Extreme Decision Making Lessons from the Antarctic* by Brad Borkan and David Hirzel

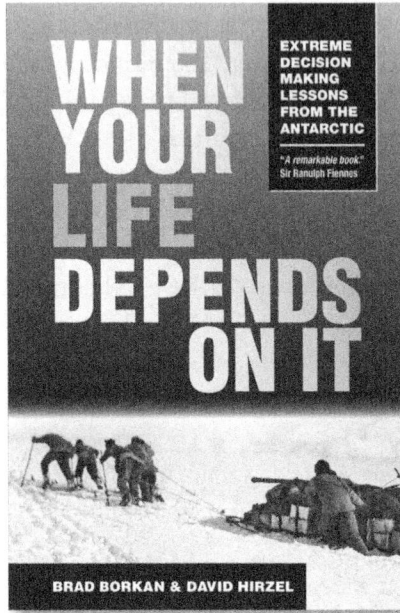

Awards

- Winner: First Place: Chanticleer International Book Awards for Insightful Non-fiction
- Winner: Wishing Shelf Awards : Best Audiobook
- Reached #19 on Top 100 Best Decision Making Books of All Time
- Finalist: Voice Arts Awards: Best Audiobook – History category.

Endorsements

"A remarkable book"
Sir Ranulph Fiennes, world's greatest living polar explorer

"Polar book of the year"
Jonathan Shackleton, Antarctic historian

It's Your Call

Antarctica – the early 1900's.
The only communication is as far as you can shout.

You and your two companions are nearing the end of a fifteen-hundred-mile trek to a nameless spot on the South Polar Plateau.

To say conditions are harsh would be an understatement. Temperatures can get so low that you risk frostbite even when bundled in your reindeer-hide sleeping bags. The jagged, frozen landscape provides constant challenges, including the danger of crevasses cracking open unexpectedly beneath your feet, plunging you into their depths. At times you have been on the verge of starvation.

Your presence here today is the result of countless decisions great and small made along the way. Right now you are faced with a decision greater than any that came before. One of your companions has fallen so ill with scurvy he can no longer walk.

Seventy miles of dangerous terrain lie ahead before you reach the safety of your base camp, and you will have to drag him on the sledge, adding an almost unbearable weight to that of your ice-encrusted tent and the last remnants of food keeping you alive.

The reality of the situation is grim. You must maintain a steady pace each day, regardless of the weather, to reach the next depot of supplies before those on hand run out. Your daily distances have fallen off, and continue to fall. The sick man, already perilously near death, is unlikely to survive the remainder of the journey.

With his extra weight further reducing your daily mileage, neither will you and your other companion. You all know the fate that lies ahead. The sick man tells the two of you to leave him here on the Barrier and march on ahead with the sledge and supplies, to save yourselves while you can. The three of you have developed a close camaraderie during your long walk; leaving him to perish on the ice is inconceivable. The obvious, ethical, human decision: to shoulder your burden and do your best.

The situation is not so straightforward. You are seamen and the sick man is your commanding officer. He has commanded you to leave him behind. The one thing that has been repeatedly drilled into you throughout your entire working life is this: there is no occasion on which you can refuse to comply with the order of an officer.

To obey means the two of you have at least a chance at survival; to refuse is mutiny, and certain death for all three of you.

The choice is now yours – it's your call. How will you decide?

★

This was a real event faced by real people. They did have to make this call. Their decision and the outcome may surprise you …

You will find the rest of the story in *When Your Life Depends on It: Extreme Decision Making Lessons from the Antarctic* by Brad Borkan and David Hirzel. It is available on Amazon and other online booksellers.

The audiobook, narrated by the world-renowned voice actor Dennis Kleinman, is available on Audible, iTunes, Spotify and all other audiobook sites.

★